D0713545

Manual of Aquatic Sediment Sampling

Manual of Aquatic Sediment Sampling

Alena Mudroch
José M. Azcue

LEWIS PUBLISHERS
Boca Raton Ann Arbor London Tokyo

Library of Congress Cataloging-in-Publication Data

Mudroch, Alena
Manual of aquatic sediment sampling / Alena Mudroch, José Azcue.
p. cm.
Includes bibliographical references and index.
ISBN 1-56670-029-9
1. Marine sediments--Sampling. 2. Lake sediments--Sampling.
3. Pollutants--Testing. I. Azcue, José. II. Title.
GC380.2.S28M83 1995
628.1′61—dc20 94-26847
CIP

Direct all inquiries to CRC Press, Inc., 2000 Corporate Blvd., N.W., Boca Raton, Florida 33431.

Lewis Publishers is an imprint of CRC Press

International Standard Book Number 1-56670-029-9
Library of Congress Card Number 94-26847
Printed in the United States of America 2 3 4 5 6 7 8 9 0
Printed on acid-free paper

THE AUTHORS

Alena Mudroch, M.Sc., is a research scientist with the National Research Institute, Environment Canada, Burlington, Ontario. Mrs. Mudroch graduated with a diploma from the Chemistry Department, State College, Prague, Czech Republic, and obtained her M.Sc. degree in 1974 from the Department of Geology, McMaster University, Hamilton, Ontario.

Currently, Mrs. Mudroch is a member of the Canadian Association on Water Pollution Research and Control, and of the International Association for Great Lakes Research (IAGLR) where she served as President (1989–90). She holds membership on several national and international committees, and is the Chief of the Sediment Remediation Water Interactions Project, Aquatic Ecosystem Restoration Branch, National Water Research Institute. Mrs. Mudroch has over 100 scientific papers and reports published, and presented over 50 papers at national and international conferences and workshops. Her current major research interests include characterizing aquatic sediments and defining the role of the pathways, fate, and effects of sediment-associated contaminants in aquatic ecosystems.

José M. Azcue, Ph.D., is a research scientist with the National Research Institute, Environment Canada, Burlington, Ontario. In 1981, Dr. Azcue graduated from the University of Basque Country in Bilbao, Spain, with a B.Sc. degree in biology and ecology. In 1983, he received a diploma of Technical Agronomist from the Ministry of Agronomy, Spain. He obtained his M.Sc. degree in biophysics in 1987, from the Federal University of Rio de Janeiro (FURJ), Brazil and his Ph.D. degree in geochemistry in 1992, from the University of Waterloo, Canada (research on the mobility of arsenic in abandoned mine tailings).

During 1990 and 1994, he lectured at FURJ, where he co-organized an international course entitled "Sampling of Aquatic Environments". Among other publications, Dr. Azcue is co-author of the book *Metales en Sistemas Biologicos*, published in 1992 by the University of Barcelona, Spain. His current major research interests include geochemical cycling and analytical determination of contaminants in the environment, industrial and residential waste treatment, technology transfer, and environmental education.

PREFACE

Aquatic sediments have been studied for many years to characterize their nature and properties for different purposes. Recently, contaminated sediments have become one of the most important environmental issues. With increasing interest in sediments acting as a sink and source of contaminants in aquatic ecosystems, it has been recognized that sediment sampling is a complex and poorly documented discipline. However, the proper collection of sediments is of great importance in environmental studies and assessment of contaminated sediments. To process nonrepresentative or incorrectly collected or stored sediment samples may lead to a waste of money and human efforts, and to erroneous conclusions.

We hope this book, designed as a field manual and general reference, will provide information to technical personnel, field managers, students, and the scientific community in general. Described techniques for sediment and sediment pore water sampling are based on years of experience of many scientists. However, this book does not discuss the theory of the scientific principles of physicochemical processes in sediments and at the sediment/water interface or analytical procedures for sediment analysis. Advantages and limitations of the different techniques and equipment for sampling are pointed out, allowing the reader to choose the most adequate sampling technique and equipment to achieve the objectives of the sediment study. One of the main goals is to discuss measures that can be taken to improve the sampling techniques and, consequently, the quality of results obtained from the sediment sampling program.

Alena Mudroch
José M. Azcue

ACKNOWLEDGMENTS

This manual of aquatic sediment sampling would not have been possible without the previous work of many individuals, both in the private sector and government agencies. We would like to express our thanks to our colleagues at the National Water Research Institute (NWRI), Burlington, Ontario, Canada, for their suggestions, help, and support. The field work experience of the personnel of the Technical Operations Section and the excellent contribution of Graphic Arts Unit members of NWRI are also gratefully acknowledged. The authors give special recognition and greatly appreciate the editorial assistance of Dianne Crabtree, and the manuscript preparation by Kate Lantagne, also of NWRI.

DISCLAIMER

Information regarding equipment, mention of trade names, or commercial products is provided as a guide and should not be construed as an endorsement of any particular device or product.

ABBREVIATIONS

AAS	atomic absorption spectrophotometry
BOD	biochemical oxygen demand
CAB	cellulose acetate butyrate
CEC	cation exchange capacity
CN	cyanides
COD	chemical oxygen demand
DDW	double-distilled water
DOM	dissolved organic matter
EPF	electromagnetic position fixing
ICP	inductively coupled plasma
I.D.	inside diameter
O.D.	outside diameter
OM	organic matter
PAH	polynuclear aromatic hydrocarbons
PCB	polychlorinated biphenyls
PIXE	proton-induced X-ray emission
PTFE	polytetrafluoroethylene (Teflon)
PVC	polyvinylchloride
QA/QC	quality assurance/quality control
rpm	revolutions per minute
RSD	relative standard deviation
TKN	total Kjeldahl nitrogen
TOC	total organic carbon
UTM	Universal Transverse Mercator (grid)

CONTENTS

Chapter 1

Introduction

Sediment sampling has played an instrumental role in many scientific disciplines. The sediment sampling techniques employed today are the synthesis of the contribution of workers from many scientific fields, such as hydrology, engineering, geology, toxicology, and oceanography. In this respect we can affirm that the roots of modern sediment sampling go back to antiquity.

In ancient times man was primarily interested in controlling nature. Only later, during the Hellenic Civilization (around 600 B.C.), did he try to understand nature (Biswas, 1970). The earliest civilizations were located along banks of rivers: the Tigris and Euphrates in Mesopotamia, the Nile in Egypt, the Indus in India, and the Huang-Ho in China. Consequently, hydrology and geology were two of the first disciplines mastered by our predecessors. One of the first tools described in the literature was the "nilometer" (Morgan, 1914). As the name indicates, nilometers were used to measure the water levels of the Nile. A series of conduits were used to bring water from the river to a well or cistern. Water levels of the Nile can be traced back to between 3500 and 3000 B.C. (Biswas, 1970).

Sedimentation, erosion, and denudation processes were first described by Aristotle (*Meteorologica* 1.14): "It is therefore clear that as time is infinite and the universe eternal that neither Tanaïs nor Nile always flowed but the place whence they flow was once dry: for their action has an end whereas time has none. And the same may be said in truth about other rivers." Strabo (*The Geography* 10.2.19 cap.458) also described sedimentation due to the silting action of rivers: "The soil is not only friable and crumbly but it is also full of salts and easy to burn out. And perhaps the meander is winding for this reason, because the stream often changes its course, and carrying down much silt, adds the silt at different times to different parts of the shore; however, it forcibly thrusts a part of the silt out of the high sea." Perhaps the recognition and interest in material brought by the streams in ancient times was the first step to the modern investigation of the character of bottom sediments.

Mining, more specifically prospecting, represented a considerable driving force for the development of new tools or principles applicable to sediment sampling. The earliest metal worked by humans was gold, followed closely by native copper and tin. Ancient tools were generally made of iron, copper, or bronze, although Roman miners sometimes used stone hammers (Shepherd, 1993). All tools appear to have had short wooden handles to facilitate their use (Sandars, 1905). The earliest method of obtaining supplies was by washing alluvial deposits. The tools occasionally found were mostly of wood, and an oaken shovel, used in about 1500 B.C., is very similar to one used in 1500 A.D. (Singer et al., 1956). Among the tools described in a copper-mine at Mitterberg (Austria) in the seventeenth century were bronze hammers to crush the ore and sieves of wooden riddles, with hazel-twigs to form the meshes (Singer et al., 1956). The ground material was then shovelled into leather bags to be transported out of the mine and carried to the valley. In general, the development of new tools for mining and prospecting was a very slow process. The functional shapes, invented during the Stone Age, remained; the materials, from which the tools were made, changed with the centuries (Forbes, 1963).

Sampling of stream and lake sediments has been used for many decades in reconnaissance or detail surveys in geochemical prospecting. The principles of geochemical prospecting are as old as man's first use of metals (Rose et al., 1979). The early prospectors learned that small fragments of fresh or weathered ore could be found in the sediments deposited by streams draining a mineralized area. However, they were guided only by what they could see with the unaided eye. The visual examination was later supplemented by analysis of different trace elements in collected stream and lake samples. Recent recovery of oil, gas, different metals, heavy minerals, and other materials from the continental shelf and ocean floor required development of various, and often relatively expensive, procedures for bottom sediment sampling to locate and outline the deposits of these materials.

Extensive sediment sampling has been carried out in fundamental studies of sedimentary and geochemical processes in marine and freshwater environments. A variety of sediment samplers have been developed and modified to be used for specific objectives and for different operating conditions in these studies. With the increasing focus on environmental issues, many projects have been carried out with the objective to evaluate the effects of contaminated sediments on water quality and/or biota in freshwater and marine systems and to remediate contaminated sediments. These studies require a well-designed program for sediment sampling. Dredging of sediments from navigation channels and harbors requires, in most cases, sampling of the sediment before dredging with the objective of evaluating its quality and selecting a proper disposal of the dredged material.

Most of the sediment sampling used in the evaluation of sediment quality is basically identical to that developed in earlier studies of sedimentary and geochemical processes in marine and freshwater environments. However, new samplers were designed and modifications were made to some old ones to achieve the quality of samples collected for evaluation of the effects of contaminated sediments on aquatic ecosystems. Sediment pore water sampling is a relatively new technique used in the identification of equilibrium reactions between minerals and water and early diagenetic mineralogical and chemical changes within the sediments. Recently, sediment pore water has been collected to evaluate processes and impacts of the transport of contaminants across the sediment/water interface into overlying water, and evaluate the effects of contaminants in the pore water on biota living in the sediments.

There is a large gap between the first scoop of "silting material" collected from the ancient rivers and the sophisticated sediment samplers and sampling techniques presently used. This book was prepared as a guide for the sampling of bottom sediments and sediment pore water in environmental studies. It should be pointed out that a proper collection of sediment and sediment pore water samples requires much more than the modern sophisticated sampling equipment. Therefore, this book contains information the authors consider necessary for the successful accomplishment of a sediment sampling program.

REFERENCES

Aristotle, Meteorologica I. XIV., Translated by E.W. Webster, Encyclopedia Britannica Inc., Chicago, 1952, 461.

Biswas, A.K., *History of Hydrology,* North-Holland/American Elsevier, 1970, 336.

Forbes, R.J., *Studies in Ancient Technology,* Vol. 1, E.J. Brill, Leider, Netherlands, 1963, 194.

Morgan, M.H., *Vistruvius, Ten Books on Architecture,* Translated by M.H. Morgan, Book 1., Harvard University Press, Cambridge, 1914, 13.

Rose, A.W., Hawkes, H.E., and Webb, J.S., *Geochemistry in Mineral Exploration,* 2nd ed., Academic Press, London, 1979, 657.

Sandars, H., The Linares bas-relief and Roman mining operations in Baetica, *Archaelogia,* 59, 311, 1905.

Shepherd, R., *Ancient Mining,* Elsevier Science Publishers, London, 1993, 494.

Singer, C., Holmyard, E.J., and Hall, A.R., *A History of Technology,* Vol. 1, Oxford University Press, London, 1956, 827.

Strabo, *The Geography* (X.2.19. cap 458), Translated by H.L. Jones, L.C.L., 1923.

CHAPTER 2

Procedures in Preparation of Sediment Sampling Program

2.1 INTRODUCTION

Extensive and thorough planning, with consideration for different scenarios, is very important for the successful completion of a sediment sampling program. Clearly stated objectives are an essential and basic requirement before planning and adequate design of the sediment sampling program can proceed. For example, the sediment sampling can represent a small part of a multidisciplinary investigation focused on environmental impacts, such as effects of industrial effluents on an aquatic ecosystem. Further, it can be part of a geological/geotechnical survey of a river, lake, or ocean floor prior to construction or oil/mineral exploration. Monitoring of sediment contaminants and assessment of sediment quality are usually carried out with the objectives to determine the extent to which the sediments are either a source or a sink for contaminants and to evaluate the effects of these contaminants on the environment of the investigated water body. Such studies can either have regulatory implications, such as dredging and disposal of the dredged material and remediation of the contaminated area, or be carried out to assess risk to human and environmental health through research of different sediment/water interaction processes.

The cost of analyzing collected sediment samples usually exceeds that of collecting them. However, the funds for the analysis are wasted if the samples are collected at inappropriate locations or do not represent the study area. Further, the proper selection and use of sediment sampling equipment, sample handling, storage, and transport are all equally important to the selection of sampling locations. Therefore, about 60% of the time allocated to the sediment sampling should be spent on detailed planning of where and how to collect the samples, including logistics associated with the travel of personnel involved in the sampling, shipping the equipment to the sampling location, and handling, preservation, storage, and transport of collected samples.

2.2 REVIEW OF DIFFERENT OBJECTIVES

The objectives of different studies of sediments are not always sufficiently described in papers published in scientific journals and government or private sector reports. However, they are extremely important in planning a sampling program.

The application of sediment research to understand environmental problems has achieved considerable success. Fundamental studies of sedimentary and geochemical processes carried out under different objectives by sedimentologists or geochemists have provided a large base for applied environmental studies. These fundamental studies include, for example, survey of distribution of types of bottom sediments within a water body accompanied by determination of sediment texture. As discussed below, understanding the distribution of bottom sediments of different textures or particle size within a water body is essential to sediment sampling for specific environmental investigations. With the increasing focus on environmental issues, many projects have been carried out with the objective of evaluating the effects of contaminated sediments on water quality and/or biota in freshwater and marine systems. Other projects have been initiated with the objective to remediate contaminated sediments either *in situ* or by removal followed by different treatments. Dredging of sediments from navigational channels and harbors requires, in most cases, sampling of the sediments before dredging, with the objective to evaluate their quality and select the proper disposal.

In some cases, the objectives of a study of sediment will define the location of the sampling stations. For example, a simple objective of a study is to determine the presence or absence of a specific contaminant in bottom sediment at a given area. In such a study, sediment can be sampled at one or a few sampling stations at fine-grained sediment deposition sites to determine the presence of the contaminant. However, after confirmation of the presence of the contaminant in the sediment, the study may be expanded to determine the extent of sediment contamination by the specific compound or element within the area, the contaminant's sources, history of the loading of the contaminant, its transport, bioaccumulation, etc. In this case, the objectives of the study become complex and need to be clearly stated before selection of the sampling stations. The objective of a baseline sediment quality survey is to determine sediment quality within a water body at a fixed point in time against which future surveys may be compared. The monitoring survey, similar to the baseline survey, involves regular or periodic resampling of sediments. However, the objective of a monitoring survey is to determine the changes in sediment quality over a period of time. For both the baseline and monitoring sediment surveys, the sediments always need to be sampled from areas of permanent accumulation of fine-grained sediments. For example, one extensive survey by the Geological Survey Canada determined the distribution of different metals and trace elements in sediments in thousands of Canadian lakes, with the objective to provide information to mineral prospecting. The samples were collected in the deepest part of the lakes, which is in most cases the area of permanent accumulation of fine-grained sediments.

Generally, the projects listed above have clearly stated objectives explaining why the project is carried out and what is expected from its execution. Regardless of the type of project, the objectives should clearly outline the physical extent of the project area. For example, the sediment quality in Lake Ontario will be monitored with the objective of evaluating the changes of inputs of different contaminants into the lake over the past 40 years. In this case, the project area is the whole of Lake Ontario. On the other hand, an investigation of environmental effects of industrial wastes disposed of near a river may involve a study of a large terrestrial area, river, lake, groundwater, etc. Therefore, the project objectives should clearly outline the physical extent of the investigated area.

2.3 SELECTION OF SAMPLING STATIONS

The purpose of sediment sampling is to collect, in the most efficient manner, samples that truly represent the sediment character in the study area. The selection of location and number of sampling stations in the study area can considerably affect the quality and usefulness of data from analysis of collected sediments in the evaluation of the effects of bottom sediments on the environment. Spatial heterogeneity of sediment chemistry within the study area also must be considered in selecting the location of the sampling stations. A sampling program should start by selecting the sediment sampling stations and preparing a sampling protocol that contains all details of the execution of the sediment sampling within the study area and, if desirable, in a selected control site.

The selection of sampling stations in any study and survey of bottom sediments depends on the desired objectives. In studies of sediment distribution, samples are usually collected on a regular grid within the entire water body of the study area. However, in most environmental studies, sediment sampling involves the collection of fine-grained sediments.

Many projects have been carried out to determine the distribution of sediments of different particle size and geochemistry in lakes and streams for which there was no previous information on sediment types (Allan and Ball, 1990). The objective of the projects was to map the horizontal sediment distribution and sediment geochemistry. Therefore, the sampling stations were selected to provide sufficient information for the mapping. On the other hand, the selection of sampling stations in a study carried out for sediment quality assessment within a given area requires knowledge of sediment distribution to select sampling stations within the areas of fine-grained sediment accumulation. Knowledge of the thickness of the fine-grained sediment layer is essential, particularly for the selection of sampling stations for sediment coring, i.e., for obtaining a vertical sediment profile.

In the absence of information on sediment distribution at the study area, it is essential to carry out a preliminary survey prior to selecting locations of the sampling sites, in order to define the permanent depositional areas of fine-grained sediments and their thickness. Examples of procedures for preliminary determination of fine-grained sediments deposition are given in Section 2.5.

2.4 SAMPLING OF FINE-GRAINED SEDIMENTS

Bottom sediments are naturally variable. Their physico-chemical properties change horizontally across a water body and vertically down the sediment profile. The bottom sediments of the study area should be subdivided into different groups, which are expected to be as homogeneous as possible. The first division of the sediments should be by particle size distribution, with fine-grained (<63 µm) sediments representing one group that is particularly important to the assessment of sediment quality.

It is generally accepted that fine-grained suspended and bottom sediment particles (silt and clay with particle size <63 µm) accumulate greater concentrations of contaminants (for example, Ackermann, 1980; de Groot et al., 1982; Förstner, 1982; Mudroch, 1984; Horowitz and Elrick, 1988), particularly those with low water solubility, than coarse particles (particle size >63 µm). The fine-grained particles exhibit properties suitable for different physico-chemical sorption and ion exchange of contaminants than the coarse particles. Further, fine-grained sediments support a large part of the benthic community by supplying the food in sediment organic matter associated with the fine-grained particles. Therefore, the assessment of sediment quality must be carried out on the fine-grained sediments sampled in areas of the water body where permanent accumulation of sediments is taking place.

2.4.1 Methods for Normalization of Concentrations of Contaminants in Bulk Sediment Samples

The association of contaminants with fine-grained sediments, particularly silt and clay particles <63 µm, was outlined above. Further, unless the sediments collected from all sampling stations within a study area consist of particles of similar size, the concentrations of contaminants in the sediments cannot be compared among the sampling stations. This rule applies particularly to river sediments and mapping the distribution of sediment-associated contaminants within the study area. The presence of coarse material, such as sand and other particles >63 µm in the sediments, effectively dilutes the concentrations of contaminants, particularly metals and trace elements, associated with particles <63 µm. However, it is possible to use normalization to reduce the effects of dilution of contaminated fine-grained sediments by coarse particles. A few methods are outlined below that can be used to compensate the effects of the dilution and normalize the concentrations of contaminants to the portion of the sediments that is active in the accumulation of contaminants.

1. When the objective of the study is surveillance of distribution of metal contamination in a water body, collected sediments from different sampling stations at a study area can be sieved (wet or dry) through a sieve with a mesh size 63 μm. For determining the concentrations of trace elements in the sediments, particularly metals, it is recommended to use a plastic sieve. Separated particles smaller than 63 μm are used to determine the contaminants. However, particle size fractions smaller than 63 μm are very difficult to separate in adequate quantity for direct analysis. In studies of transport and mass loadings of contaminants, analysis of a single separated particle size fraction may not be inadequate.
2. The concentrations of contaminants in a bulk sediment sample can be corrected for fine-grained sediments by using the information on particle size distribution in the sample. In this approach, it is assumed that the relationship between the contaminant, such as a trace element, and the particle size, i.e., the <63 μm particle size fraction, is linear. Further, it is necessary to carry out an accurate and precise determination of the contaminant and particle size distribution in the sediments. For example, sediments with high concentrations of organic matter, iron, or calcite may require a specific treatment prior to the determination of the particle size distribution. High concentrations of iron, calcite, or other elements and compounds can generate aggregates that are difficult to disperse by conventional methods used in the determination of the particle size distribution. As suggested by Loring and Rantala (1992), concentrations of the contaminants, such as a metal, should be plotted against the percentage of the <63 μm particle size fraction in the sediment to establish the relationship between the contaminants and grain size. They argue that in most cases some type of linear relationship emerges in the general form $y = ax + b$. A linear relationship in the form $y = ax$ is uncommon because the >63 μm particles usually contain some of the contaminants, particulary metals and trace elements. A significant relationship, indicated by a high correlation coefficient and $p \leq .05$ or .01, suggest that the contaminants' grain size normalization is suitable for the sediments collected at the study area. The limitation to this method is the requirement of a large number (10–15) of samples from a study area. Further, in analyzing the data it is often impossible to obtain a regression line, due to the limited range in particle size of the sediments at the study area.
3. The concentrations of contaminants can also be corrected by using the concentrations of a conservative element in the bulk sediment sample. For example, aluminum is a conservative element, i.e., it occurs naturally in alumino-silicates that are mainly in the silt and clay particle size fraction of the sediments. Therefore, aluminum is a surrogate of the fine particle size fraction of the sediments. In considering the use of aluminum or other conservative element in the normalization, the dependence of concentrations of aluminum or the other conservative elements in the bulk samples on the <63 μm fraction of the sediment should be established. The conservative element used in the normalization must be an important constituent of one

or more of the major fine-grained carriers of the contaminant, such as aluminium is in the clay particles, and must reflect its variation with a different sediment particle size. For example, it has been shown that lithium is superior to aluminum for the normalization of trace metal data from glacial sediments and for the identification of anomalous metal concentrations (Loring and Rantala, 1992). One of the major disadvantages of using a conservative element is that the contaminant/conservative element normalization yields a ratio value instead of a real concentration of the contaminant in the sediments. Ackermann (1980) used the ratio of concentration of heavy metals to concentration of cesium in bulk sediments collected from the Emo River estuary to calculate the concentration of heavy metals in sediment particle size ≤20 μm. He found that cesium was well correlated with the percentage of ≤20 μm particles in collected sediments (correlation coefficient $r = .987$, sample number $n > 70$), and anthropogenic heavy metals were mainly associated with the ≤20 μm sediment particles. Further, contribution of cesium by the coarser-grained sediment fractions changed inconsiderably the ratio. The concentrations in the ≤20 μm particle size fraction was calculated by

$$C_{HM}(\text{fraction } \leq 20\mu\text{m}) = \frac{DC_{HM}(\text{bulk sample})}{C_{cesium}(\text{bulk sample})} \times F$$

where C_{HM} and C_{cesium} are the concentrations of heavy metal and cesium, respectively, measured in mg/g sediment dry weight, and F is a conversion factor derived for the extraction of a regression line of concentration of cesium plotted against the percentage of ≤20 mm particle size fraction in bulk sediment samples (Figure 2-1).

Organic contaminants are usually associated with the organic matter in the sediments and the standardization should be made using the content of organic matter in the bulk sediment sample. However, in most cases there is a good

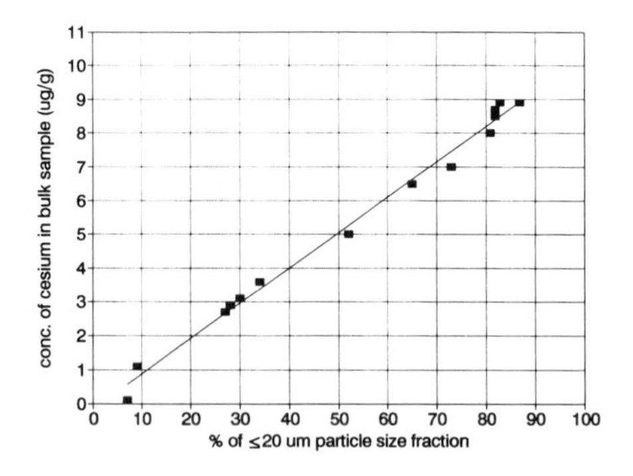

Figure 2-1 Concentration of cesium in bulk sediment sample as a function of the particle size fraction ≤20 μm (after Ackermann, 1980).

correlation between the concentration of organic matter and the amount of fine-grained particles in the sediments. Therefore, the above-described corrections using the particle size and concentrations of aluminum can also be applied for the correction of the concentrations of organic contaminants in sediment samples collected at different sampling stations within a study area. Other conservative or inert constituents of the sediment, such as cobalt and scandium, or a metal-reactive component of the sediment that is not affected by anthropogenic activities, can be used in normalizing contaminant concentrations in bulk sediment samples collected in different areas. However, it was suggested by Horowitz and Elrick (1988) that normalization procedures are useful for clarifying trends but often over- or underestimate metal concentrations. On the other hand, physical separation of fine-grained from coarse sediment particles followed by chemical analysis clarifies trends and produces accurate measurements of metal concentrations. The use of the particle size and concentrations of different conservative elements in normalization requires a collection of large amount of analytical data. Further, the sediments within the study area must be derived from the same material sources, and anomalous metal levels must be known.

2.4.2 Location of Fine-Grained Sediments in the Study Area

Fine-grained sediments are not uniformly distributed in lakes, oceans and, particularly, rivers. The distribution of the sediments on the lake, river, or ocean floor is affected by energy-controlled processes. The sorting of sediments by texture or particle size is a hydraulic process with fluctuating intensity and location affected mainly by current and wave action. Fine-grained sediments usually accumulate in low-energy zones, such as a bay or the outer side of the main channel of a meandering river. Coarse particles, such as sand and gravel, and hard glaciolacustrine clay are found on the bottom of fast-flowing rivers. In large lakes, such as the North American Great Lakes, the deposition of fine-grained sediments is affected by energy-controlled processes such as current circulation and wave action. The fine-grained sediments are deposited at deep, quiet waters in the offshore areas. Similarly, in small lakes, fine-grained sediments usually accumulate in the greatest water depth. However, there can be many exceptions in fine-grained sediment deposition in small lakes. Scientists involved in the selection of sampling stations in the assessment of sediment quality should have at least a basic knowledge of the distribution of fine-grained sediments at the study area. If this information is not available, a preliminary study is needed to define the fine-grained sediment deposition in the area.

In lakes and rivers of glacial origin, it is necessary to distinguish between recent and glaciolacustrine fine-grained sediments. The glaciolacustrine sediments were brought by meltwater streams flowing into lakes bordering the glacier or derived from reworking of exposed glacial tills. Glaciolacustrine sediments are typically very fine, firm, and sticky clay-size materials containing about 25% water, in contrast to the fine, soft recent sediments containing about 70 to 95% water. The consistency of the glaciolacustrine sediments prevents the penetration of grab samplers and most of the sediment corers. These sediments, usually found in a high-energy zone of a lake or river lacking deposition of recent sediments, do not contain anthropogenic contaminants. In some lakes they underlie the recent sediments.

2.4.3 Heterogeneity of Fine-Grained Sediments

In addition to the nonuniform distribution of fine-grained sediments in lakes, rivers, and oceans, many factors contribute to physico-chemical heterogeneity of fine-grained sediments in depositional areas. These factors include, for example, the distance from a pollution source or a stream delta, bathymetry of the water body, and sediment resuspension and redeposition. Several studies indicated great variation in sediment chemistry, even within small sampling areas with relatively uniform distribution of fine-grained sediments (Dowring and Rath, 1988). Therefore, the selection of the sampling stations is complicated by the great variety of factors that can affect the spatial heterogeneity of fine-grained sediments within a water body. The complexity of this problem extends further as the factors vary from one water body to another. Consequently, local and detailed information on the characteristics of the water body to be sampled is necessary in the selection of the sampling stations.

Following are examples of the heterogeneity of fine-grained (<63 μm) sediments in different water bodies.

Sediment Heterogeneity in a Large River

A survey of bottom sediment quality was carried out in 1991 at the Cornwall area in the St. Lawrence River, Canada (Mudroch, P., 1993). The St. Lawrence River, with a flow rate of 7,400 m^3/s and length of 450 km, drains the Great Lakes into the Gulf of St. Lawrence of the Atlantic Ocean. Many contaminants are transported into the river from the Great Lakes and many others from point and diffuse sources along the river. The river has a complex morphometry, with many channels, islands and riverine lakes, and a dredged navigational channel maintained at approximately 9 m water depth. Despite the fast flow, there are several areas with fine-grained sediment deposition on the river bottom. The objective of the sediment survey at Cornwall was to provide information in planning of remediation of contaminated sediments at the area. Fine-grained bottom sediments were collected within two areas determined by a side scan sonar/subbottom profiler system as depositional zones of fine-grained sediments. Surface sediments were collected at eight sampling stations within these zones using a Ponar grab sampler (Figure 2-2). At stations 3 and 10, three grab samples were obtained to determine the heterogeneity of the fine-grained sediments in the study area. Each of the six grab samples was further divided into three subsamples, and each of the eighteen subsamples obtained was treated and analyzed as an

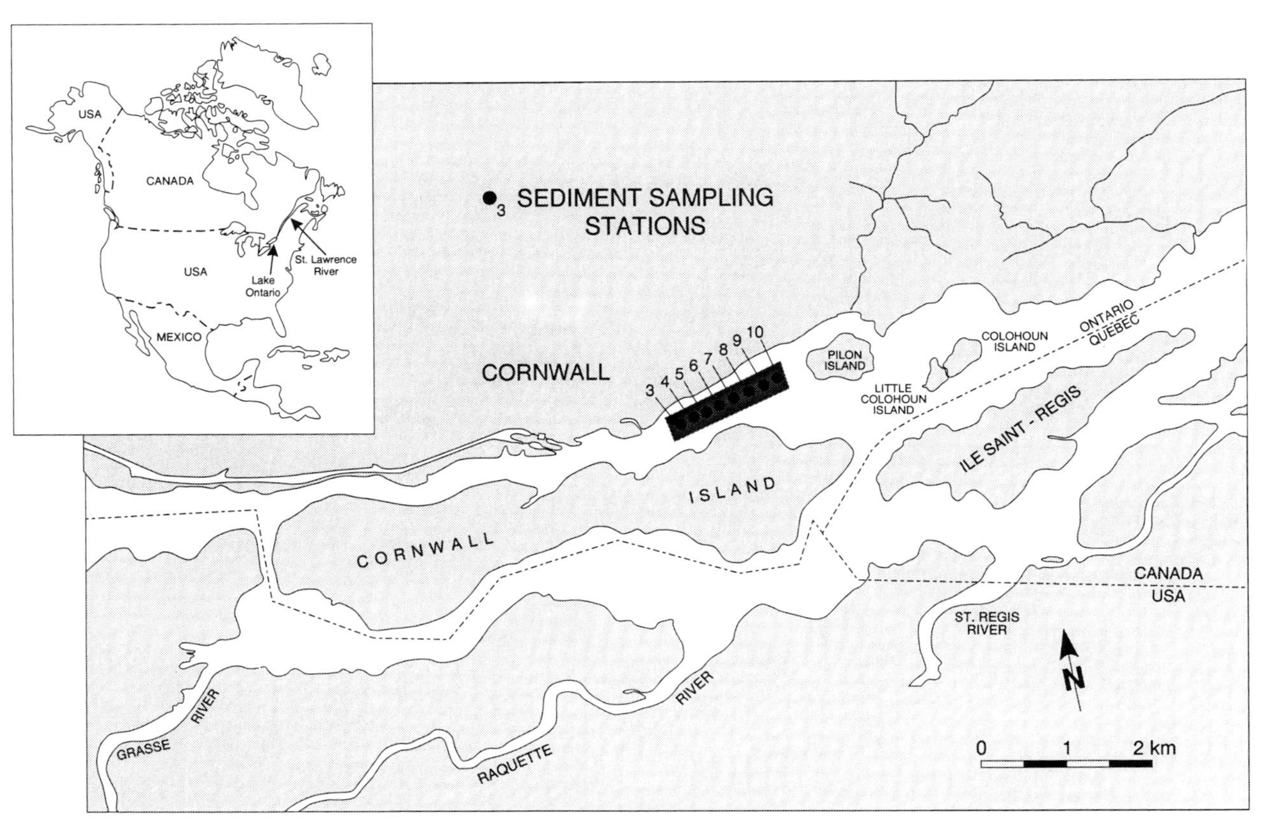

Figure 2-2 Sediment sampling stations in the St. Lawrence River at Cornwall.

individual sample. The heterogeneity of the sediments was evaluated using concentrations of selected major and trace elements and organic contaminants in each sample. The example of the results (Table 2-1) shows the concentrations of four trace elements in each of the eighteen subsamples and their relative standard deviation within one Ponar grab. It was found that the concentration of the trace elements varied up to 81% relative standard deviation (for cadmium, not shown in Table 2-1) within one Ponar grab. The variation of the concentrations of some elements in the sediment was within the accuracy of the analytical method used in the laboratory to determine the elements (Table 2-2). Greater heterogeneity occurred at station 3, particularly within the first Ponar grab, which was located closer to the point source on the shore than station 10. The variation in concentrations of elements was different for each element and represented the heterogeneity of the fine-grained sediments in the area.

Sediment Heterogeneity in a Large Lake

A sediment sampling program was designed to study heterogeneity of fine-grained sediments in the western, central, and eastern depositional basins of

Table 2-1 Sediment Heterogeneity in the St. Lawrence River at Cornwall (concentrations in µg/g dry weight)

	Cu	Pb	Zn	Hg
Station No. 3				
1st Ponar				
Subsample 1	25	17	63	6.03
Subsample 2	16	5	49	2.70
Subsample 3	20	10	54	4.00
RSD[a]	22%	56%	13%	40%
2nd Ponar				
Subsample 1	18	12	70	0.73
Subsample 2	17	13	64	0.80
Subsample 3	20	10	66	0.78
RSD	8%	13%	4%	5%
3th Ponar				
Subsample 1	18	16	68	1.12
Subsample 2	14	15	60	0.90
Subsample 3	16	12	63	1.12
RSD	13%	15%	6%	15%
Station No. 10				
1st Ponar				
Subsample 1	34	34	301	1.06
Subsample 2	35	35	311	1.42
Subsample 3	32	32	281	1.36
RSD	4%	6%	5%	15%
2nd Ponar				
Subsample 1	34	33	277	1.32
Subsample 2	35	30	324	1.11
Subsample 3	34	30	286	1.14
RSD	3%	6%	8%	9%
3th Ponar				
Subsample 1	33	30	309	1.02
Subsample 2	35	35	317	1.17
Subsample 3	34	34	329	1.38
RSD	3%	9%	3%	15%

[a] RSD = relative standard deviation.

Table 2-2 Accuracy of Analysis of Sediments Collected in the St. Lawrence River at Cornwall (concentrations in μg/g dry weight)

Element	Certified standard value	Reported value
Copper	25.1 (RSD 15%)[a]	24
Lead	34 (RSD 18%)	32
Zinc	191 (RSD 8.9%)	203
Mercury	4.57 (RSD 5%)	4.61

[a] RSD = relative standard deviation

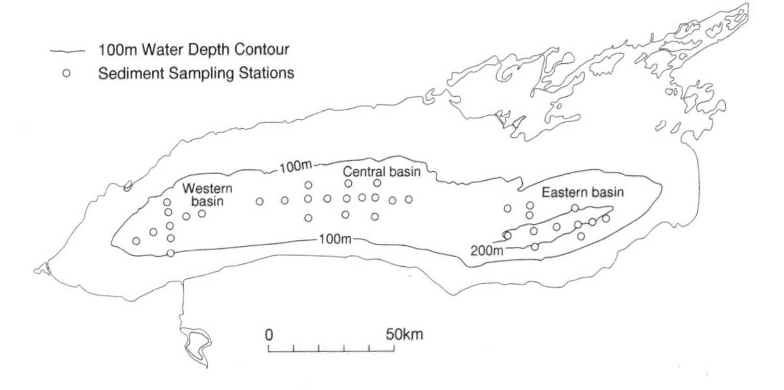

Figure 2-3 Sediment sampling stations in Lake Ontario.

Table 2-3 Sediment Heterogeneity Within a Single Box Core Collected in Lake Ontario (concentrations in μg/g dry weight)

Element	Range	RSD (%)[a]	RSD (%)[b]
Copper	105–113	2.8	1.9
Lead	125–133	2.3	1.8
Zinc	394–419	1.9	1.1
Nickel	83–85	1.1	1.1

[a] Relative standard deviation calculated from analyses of nine fields within the box core.

[b] Relative standard deviation calculated from nine repeated analyses of one sample from the box core.

Lake Ontario, one of the North American Great Lakes (Mudroch, A., 1993). The results of the study were used to recommend long-term monitoring of contaminant inputs into Lake Ontario using fine-grained sediments as a historical record of pollution. Surface sediments and sediment cores were obtained at 36 sampling stations located along transects within each depositional basin (Figure 2-3). All sediment samples were collected using a box corer of size 0.5 m × 0.5 m × 0.5 m. The surface 3 cm of the sediments were subsampled from each box core and analyzed for major and trace elements. In addition, the heterogeneity of the surficial sediments in a single box core was investigated by dividing one box core into nine equally sized fields, and each field was treated as an individual sample. The heterogeneity was evaluated using con-

Table 2-4 Heterogeneity of Sediments in Lake Ontario Depositional Basins: Concentrations of Trace Elements in Surficial Sediments (in µg/g dry weight)

	Range	RSD (%)[a]
Copper		
Western basin	108–120	3.2
Central basin	93–120	9.0
Eastern basin	93–128	9.9
Lead		
Western basin	120–144	6.8
Central basin	135–185	10.7
Eastern basin	107–171	14.7
Zinc		
Western basin	360–450	9.5
Central basin	270–390	9.5
Eastern basin	250–370	10.0
Nickel		
Western basin	77–93	6.3
Central basin	67–89	7.4
Eastern basin	63–77	8.6

[a] RSD = relative standard deviation.

centrations of major and trace elements in the sediments. It was found that the heterogeneity of the sediment within a single box core did not affect the comparison of the heterogeneity of sediments within and among the three Lake Ontario depositional basins (Table 2-3). The concentrations of selected trace elements in surficial sediments from the three Lake Ontario depositional basins are shown in Table 2-4. The results of the study indicated relatively homogeneous distribution of fine-grained sediments in Lake Ontario.

For more details on the processes characterizing the bottom dynamics in a lake and surficial sediment distribution in relation to energy-controlled processes, we suggest reading Sly and Thomas (1974), Håkanson (1977), and Baudo (1990).

2.5 PROCEDURES FOR A PRELIMINARY DETERMINATION OF FINE-GRAINED SEDIMENTS DEPOSITION

To correctly select the location of sediment sampling stations in studies of sediment contamination, it is necessary to obtain information on the type of sediments, particularly on the location of fine-grained sediments and their extent at the study area. Generally, two methods are available to obtain such information. The first is an acoustic survey of the bottom of the water body to be sampled and the second is limited-scale sediment sampling at selected locations. Preliminary information obtained by one of these methods or a combination of both will provide guidance in the design of appropriate selection of sediment sampling stations in the final sampling program.

Acoustic survey techniques such as echo sounding, seismic reflections, and refraction are able to characterize both the type of surficial sediment layer, such as sand, gravel, or soft silty clay, and the subsurface sediment layers. Acoustic penetration is most effective in unconsolidated sediments consisting of soft, silty clays with high water content. On the other hand, sands or firm, compacted sediments display minimal acoustic penetration. Seismic reflection is based on a principle similar to echo sounding, but uses a low frequency sound and allows a deeper penetration, even into a coarse sediment. Seismic refraction is based on the lateral movement of sound between an explosive source and a sound receiver. The time of arrival of specific responses is related to the thickness of sediment layers through which the sound travels. These techniques require special equipment and a good positioning capability. A correct interpretation of records obtained by an acoustic survey often requires the help of a geophysicist.

In a preliminary survey of sediment distribution by sampling, the locations of the sampling stations generally depend on the size and shape of the water body to be sampled. In a lake, preliminary sampling stations should be located on a transect down the long axis of the lake. Additional sampling stations should be located on few transects across the lake. The minimum number of samples will depend on the bottom morphometry. For example, if the lake has a regular shape and bathymetry, the preliminary sampling stations will be less dense toward the center of the lake (Figure 2-4). In irregularly shaped lakes with a variable water depth, more transects will be necessary. At a river mouth or near a point source, sampling stations can be located on a ray-shaped transect. In a study area with a system of different water bodies, such as

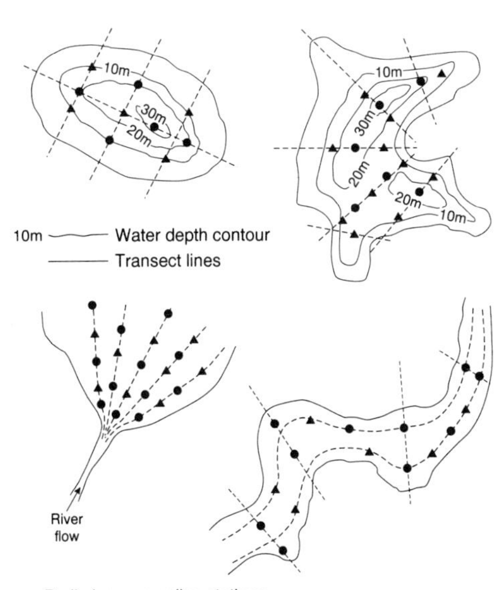

Figure 2-4 Design of preliminary sediment sampling in different water bodies.

streams, and small and large lakes, preliminary sampling stations can be located at the deepest point of the lakes where, in most cases, fine-grained sediments accumulate. For the selection of the preliminary sampling stations in a stream, recommendations given in Section 2.6 should be considered. Transect lines for locating sampling stations can run along or across the river, according to the water depth and morphometry of the river, particularly in a meandering river (Figure 2-4). There is usually very little deposition of fine-grained sediments on the bottom of a fast-flowing river.

The spacing of the sampling stations along the transects depends on available time, funds, etc. However, after collection of the samples and their visual inspection, additional sampling stations can be considered along the transect to obtain more information. This approach is illustrated in Figure 2-4 showing the location of preliminary sampling stations along chosen transects and additional sampling stations where more information may be required.

At the preliminary sampling stations, information on the physical characteristics of sediments may be obtained by the fall cone technique described by Håkanson and Jansson (1983). A sediment penetrometer, for *in situ* measurement of sediments' physical characteristics, can rather simply and rapidly determine sediment types and prevailing bottom dynamics. The penetrometer has a few different cones. The depth of the penetration of these cones reveals the nature of the sediments. A calibration table is supplied with commercially sold sediment penetrometers to convert the penetration depth into several different sediment types ranging from very soft to hard bottom sediments.

Representative sediment samples should be obtained at each of the selected preliminary sampling stations for visual inspection. The type of sediments and water depth should be recorded for each of the stations. By using a combination of acoustic survey, description of sediment texture, and a bathymetric chart, the distribution of different types of sediments can be predicted. This information is further used in the preparation of the final sediment sampling plan, including location and number of sampling stations, list of sampling equipment, logistics of the field work, quantity of sediments to be sampled, etc.

2.6 SELECTION OF SAMPLING STATIONS IN A RIVER

Bottom sediment sampling has been applied more to lakes than rivers. However, sampling techniques used in lakes can be applied in rivers. There are several conditions that must be considered in selecting bottom sediment sampling stations in a river. Rivers with a high-speed flow have few areas of fine-grained sediment accumulation. Further, some of these depositional areas are only transitional, i.e., the fine-grained particles settle on the bottom and are subjected to subsequent resuspension and transport into another downstream depositional area. The selection of sampling stations for collecting fine-grained sediments in fast-flowing rivers is limited mainly to natural sediment traps, such as the inside of channel bends, isolated pockets along the margins of the channel, areas sheltered from the main flow, etc. In rivers with a moderate or slow flow (mainly meandering rivers), a preliminary sediment sampling along transects across the river will reveal the extent of deposition of fine-grained sediments on the bottom. The bottom sediments in a river are subjected to continuous changes in flow and discharge. Seasonal variations within sediment depositional areas, such as more intensive erosion and transport of fine-grained bottom sediments during the spring and fall runoff, need to be considered.

Due to the conditions described above, the selection of bottom sediment sampling stations in rivers is more difficult than in lakes where depositional areas of fine-grained sediments generally remain unchanged.

Following is a suggested sequence of steps in the selection of sediment sampling stations in a river:

1. Carry out a physical inspection of the area to be sampled prior to selecting the sampling sites.
2. Consider sampling in areas sheltered from the main flow.
3. Sample under low flow when most of the fine-grained suspended sediments become deposited on the river floor.
4. To determine fine-grained sediment depositional areas (given by the objectives for bottom sediment sampling), carry out a preliminary sediment sampling along one or a few transects across the river. The number of sampling points along the transect(s) should be based on the flow speed (more points at the areas with low flow speed), water depth, and the width of the river.

Experience and visual inspection of the site together with knowledge of riverine environments are of great advantage in the selection of sampling stations in a river.

2.7 REQUIRED NUMBER OF SAMPLING STATIONS

At a study area with some available information on the quality of bottom sediments, the first step in selecting the location and number of sampling stations should be the estimation of variability and mean concentration of chemicals of interest in the sediments. This estimation should be followed by a statistical method to calculate the number of samples needed to achieve the acceptable confidence levels and to meet the objectives of the sediment sampling program.

When there is no information available on sediment chemistry at the study area, the number of required sampling stations can usually only be estimated after statistical analysis of a preliminary sampling program. The preliminary program should determine the heterogeneity of sediment chemistry and concentration ranges of the chemicals of interest in the study area. The number of samples theoretically required is often unrealistic, and technical and economical constrains have to be considered in the selection of the number of sampling stations in the study area. Consequently, the use of additional statistical analysis may be helpful after determining concentration ranges of chemicals of concern in the sediment through preliminary sampling. The preliminary sampling program can be very costly and time-consuming. Further, few examples are available of statistical methods applied to the selection of the location and number of sampling stations in bottom sediment sampling programs, particularly with the objective to minimize the cost or maximize the information obtained within a given budget. The extent of geochemical heterogeneity and different types of sediments are two of the many variables not considered in statistical methods. Often common sense, good judgment, knowledge of historical background of the sampling area, and experience are the best help in selecting the number and locations of the sediment sampling stations.

2.7.1 Examples of Calculation of a Necessary Number of Sediment Sampling Stations

Statistical concepts in soil sampling described by Crépin and Johnson (1993) can be applied, with small modifications, to the selection of the number of sampling stations in bottom sediments sampling. Their approach to the selection of the number of sampling stations is clear and relatively simple. The following are important factors in selection of the number of sampling stations suggested by Crépin and Johnson. The most familiar and important statistic used in soil science is the mean, $\bar{x}$, defined as the arithmetic average of a variable among a collection of samples. The variance, s^2, measures the dispersion of individual samples around the mean. A wide variance indicates wide dispersion, a small variance indicates little dispersion. The mean concentration of a contaminant in sediment calculated from three samples will most likely be different from another three samples obtained within the same area. Using the example of fine-grained sediment heterogeneity in the St. Lawrence River shown in Table 2-1, the mean concentrations of lead in the three subsamples collected from three Ponar grab samplers were 10.7, 11.7, and 14.4 µg/g at the same sampling station in the river. On the other hand, the mean concentrations of mercury in the same sediment samples were 4.24, 0.77, and 1.05 µg/g. A contaminant concentration in sediment samples ranging from 10 to 15 µg/g inspires more confidence and allows easier assessment of sediment quality than a 5 to 25 µg/g range. The standard error of the concentration, s_x, indicates the reliability of the mean. As Crépin and Johnson suggested, for a large number of samples (>50) collected at a study area, we can be 95% confident that the actual population mean will be within two standard errors of the sample mean. Therefore, for a sediment with a sample contaminant mean of 11 µg/g and a standard error of 1.5 µg/g, the true population mean concentration will be within 8 to 14 µg/g of the contaminant. This estimate will be wrong, because of natural sampling variation, only 5% of the time. The values obtained by adding or subtracting two standard errors to the sample mean are 95% confidence limits (CL). Confidence limits for any level, such as 67, 80, 90, 95, or 99%, can be calculated from the sample mean and standard error.

The cost for sampling and analyses of the sediment can be decreased by taking only as many samples as are needed for a given level of precision. The following example calculating the necessary number of sampling stations was modified after Crépin and Johnson. For example, unless a 5% chance of error occurs, we want the sample mean to be within ±1.5 μg/g of the population mean. The following formula is used in calculating the number of needed sampling stations (n):

$$n = \frac{t^2 s^2}{D^2}$$

where: t is a number chosen from a "t" table for a chosen level of precision, in this example, 95%; the degrees of freedom for t are first chosen arbitrarily, for example 10, and then modified by reiteration.

s^2 is the variance, known beforehand from other studies or estimated by $s^2 = (R/4)^2$, where R is the estimated range of concentration likely to be encountered in sampling.

D is the variability in mean concentration of contaminant we are willing to accept. (The formula is based on the assumption that the sample mean is normally distributed.)

Therefore, for a sediment with a concentration of a contaminant ranging from 0 to 13 μg/g, the estimated number of sampling stations at 95% probability and within 1.5 μg/g of the true mean CL will be

$$n = \frac{(2 \cdot 23)^2 (3 \cdot 25)^2}{(1.5)^2} = 23$$

Since 23 sampling stations are considerably more than the 10 we used to obtain a t value, we need to run the following reiteration using a t value equal to our new estimate:

$$n = \frac{(2 \cdot 069)^2 (3 \cdot 25)^2}{(1.5)^2} = 20$$

A quantitative determination of a contaminant in twenty samples can still be expensive, but the estimated variance is high, i.e., a range from 0 to 13 μg/g of the contaminant in sediment. Any sampling scheme that reduces the variance will lower the number of needed sampling stations. The number can also be lowered by relaxing the probability from 95 to 90% or by allowing the confidence limits to increase to ±2 μg/g or greater.

Table 2-5 Summary of Steps Prior to Selecting Sediment Sampling Stations

- Clearly state the study objectives.
- Clearly state the objectives and goals of the sediment sampling.
- Outline available funds, time, and effort allocated to the sediment sampling program.
- Determine the extent of the study area.
- Record all available information about the study area.
- Determine the water bodies from which the bottom sediments have to be sampled.
- Determine the size and depth of the water bodies.
- Estimate the number of sampling stations necessary to achieve the objectives of the sediment sampling and those of the study.
- Prepare a sediment sampling protocol.

In addition to considering expenses for sampling and analyses of the sediment samples, the project objectives are extremely important in the selection of the number of sampling stations. For example, an assessment of sediment quality needs to be carried out in a harbor. According to guidelines for sediment quality, the sediment in the harbor containing over 10 μg/g of the contaminant must be remediated. However, the remediation is very costly. On the other hand, the concentration of 10 μg/g must be considered when using the above method for calculating the necessary number of sediment sampling stations in the survey of sediment quality in the harbour.

A summary of steps recommended prior to selecting the sampling station is shown in Table 2-5. For further reading on the statistical approach relevant to bottom sediment sampling and analysis of obtained data, we recommend Baudo (1990) and Keith (1988). In addition, examples of the selection of sediment sampling stations are given in Section 2.10.

2.8 POSITIONING OF SAMPLING STATIONS

Sediment sampling projects, either reconnaissance surveys or detailed surveys, according to their scope, must start with marking the positions of the planned sampling stations on good-quality navigation charts in the horizontal plane in latitude and longitude, grid-coordinates or angles, and distances from known control points. Topographic maps of the adjacent shore will be necessary for some positioning techniques. The primary task is to carry out the sampling as near to these marked positions as possible.

Accurate positioning of sediment sampling sites is particularly important in programs in which the sampling is to be subsequently repeated. The required accuracy and precision of the station position will depend on the nature of the study. A 5 to 10% error in the distance between sampling sites can be acceptable for a baseline survey study carried out over a large area (more than 10 km^2). In studies involving monitoring of sediment contaminants over a certain time period, or changes over a small geographic scale, accuracy of positioning within a few meters may be necessary.

The choice of proper positioning methods depends on the project area and the distance between the sampling stations and the shore of the water body. A positioning technique suitable for the sampling program must be selected before the commencement of the sampling. Many vessels have a good positioning system and personnel familiar with its operation. When the owner or operator of the sampling vessel does not have the necessary expertise and/or positioning equipment, a competent surveyor, a survey company, and/or government agencies should be contacted for advice to assist in positioning. Regardless of the type of system selected, the positioning equipment must be properly set up and calibrated, and standard operating procedures must be followed by trained personnel to achieve maximum accuracy of the positioning system.

The distance line or taut wire can be used for direct positioning of sampling stations within a small sampling area or close to the shore. In this method, a line marked along its length is stretched between the vessel and the shore whereby the distance-off may be measured as required.

For small areas, such as harbors, embayments, small lakes, and reservoirs, where sampling is carried out from a small vessel or from ice, the sampling sites can be determined by sextant observations of structures on the shore and shoreline features in combination with navigation or topographic maps. Good visibility is necessary for this positioning technique.

Sampling of large areas, such as harbors, lakes, and oceans, with a large number of sampling sites, requires an electronic positioning system, such as Hydrotrac, Geoloc, Mini-Ranger, Trisponder, radar, etc. These systems, operating on radiolocation frequency bands and using a variety of measurement techniques, are used where other methods of positioning are not continuously available or not accurate enough. The range depends on the height of the transreceiver units, and is usually 25 to 80 km. The accuracy can be ±1–3 m. The systems are portable and relatively easy to operate. However, they require multiple onshore stations with proper security and operation, which increases their cost.

Radio signals have been used for many years to provide homing references and lines of position. There are many radionavigation systems currently in operation that have been used extensively by the civil sector. Radio methods include "line-of-sight" microwave electromagnetic position-fixing (EPF) systems. They utilize remote instruments (similar to tellurometers used in land surveys) ashore at a distance of up to 80 km. Medium-range EPF systems have a range capability of 150 to 1,200 km. Long-range navigational systems can be used from 150 km to virtually world-wide operations.

The Shoran (for short range system) operates with signal pulses in the 200 to 300 MHz band. A trigger signal from a beacon fixed on the vessel causes two transmitters at fixed shore stations to emit signals, giving two direct-range measurements when the signals are received back on the vessel. The observed time differences represent the transit time from the transmitters to the beacon and back, and are recalculated and expressed as distance. Shoran is limited in range, i.e., <80 km, although it provides high precision (±10 m at maximum range). The Distomat DI 3000 with precision ±0.1 m is suitable for harbors and small areas.

Loran C is one of the microwave electromagnetic position fixing (EPF) systems with a range up to 2,800 km. The name is derived from long-range navigation. The expanded configuration of the Loran C system is operated on a full-time basis by the U.S. Coast Guard as a service to private and commercial navigation around the North Atlantic and North Pacific oceans. It has been installed in many other areas around the world. However, it is necessary to check the sampling area for availability of reliable continuous Loran C signals. The system functions in chains, each having a master station and several secondary (slave) stations for synchronized transmission. Each pulse transmitted in the Loran C system covers a number of cycles of the 100 kHz signal frequency. A Loran master station transmits nine pulse envelopes spaced 1 m apart, then waits for a specified time before repeating its nine-pulse pattern. The various slave stations in the chain emit eight-pulse transmissions. The time difference in arrival of pulses from the master station and secondary stations gives hyperbolic lines of position. Travel pulse time must be converted to distances. The accuracy obtained with Loran C is relatively stable with time, but varies with location. Using the best techniques for calibration, accuracy of ±15–65 m can be achieved. An uncalibrated Loran C system is likely to yield errors of 500 m.

Decca and Omega navigation systems are based on similar principles. In the Decca system, phase comparison of continuous waves gives hyperbolic lines of position. The Omega system is based on phase differences between continuous waves from synchronized transmitters. The range of Decca systems is up to approximately 460 km, and accuracy is from about 1 to 60 m. Omega is a low frequency system operating at 10 kHz, providing low-order navigation signals that can be received worldwide. With a receiver, small boats and aircrafts can determine their approximate position. However, Omega is not recommended for navigation in geophysical and environmental investigations in oceans, lakes, and rivers because of its poor accuracy of several kilometers. The DECCA NAVIGATOR system is a major navigation system for air and maritime users, with facilities in western Europe, the Baltic, South Africa, the Arabian Gulf, India, Japan, and Australia (Radio Technical Commission for Maritime Services, 1990).

Satellite navigation systems (SATNAV) consist of a number of navigational artificial satellites in polar orbits and are used extensively for positioning systems worldwide. For example, with the Transit system, a fix is possible during the satellite's pass, which lasts about 15 minutes at intervals that vary from about 35 to 100 minutes depending on the observer's latitude. The satellites transmit a navigational message lasting precisely 2 minutes. Using many data and complex fix computation techniques, the accepted vessel position is calculated. The Transit satellite system was developed as an all-weather tool to provide position fixes for ships at sea, and the geophysical industry was the first among commercial users to apply satellite navigation methods to improve positioning in offshore surveys. The most upgraded system should bring the standard error of the translocated fixes within the range of 5 to 10 m. Transit coverage is worldwide. However, plans exist to terminate Transit operations in the mid-1990s. The NAVSTAR Global Position System (GPS) was developed recently. It is a satellite navigation system providing highly accurate position, i.e., ±100 m and 0.1 to 1 m for differential GPS, and velocity information in three dimensions, as well as precise time, to users around the globe 24 hours a day (Radio Technical Commission for Maritime Services, 1990). However, the continuous calculation of positions is presently not feasible for satellite navigation systems, because the time between fixes is up to 3 hours long. Consequently, it is desirable to have another navigation system available that will operate between the satellite navigation positions.

Underwater positioning methods use different types of underwater acoustic beacons at known positions on the sea or lake floor, and sensors such as echo sounders, sonars, TV cameras, etc. The beacons can also be used for marking instruments on the lake or ocean floor.

The above descriptions of positioning systems give brief information on what is available in different parts of the world. However, positioning methods are described in many books in great detail, and anyone who needs more information should consult them. Government agencies may provide information on a large-scale positioning system available within the country. Information on small-scale positioning systems may be obtained from private companies. Many low-frequency, long-range navigation systems are permanently established and operated by national and international agencies, for example, Loran C by the U.S. Coast Guard.

2.9 EXAMPLE OF A SEDIMENT SAMPLING PROTOCOL

One step that should be undertaken prior to selecting the sampling stations is the preparation of a sediment sampling protocol. An example of such protocol is given below.

1. Define the character of the samples to be collected to meet the objectives of the sediment sampling program and the study objectives. For example, in the collection of surface sediments, the depth of surface sediments to be collected, and in the collection of sediment cores, the length of the sediment cores need to be defined.
2. Confirm available funds and number and availability of trained and nontrained personnel needed for the sediment sampling program.
3. List all physical, biological, and chemical analyses (including bioassays) that will be carried out on collected sediment samples in the laboratory as well as observations and tests that will be carried out in the field.

4. Discuss and confirm with the laboratories a list of individual analyses and assays of sediments collected in the study area. Estimate and compile the quantity (i.e., volume, weight) of wet and dry samples necessary to carry out all listed analyses and assays. In the estimation, consider analyzing duplicate samples, banking collected sediment samples for future analysis, and performing other QA/QC procedures that will require an additional volume or weight of samples.
5. Collect information on various parameters in the study area relevant to the sediment sampling program, such as water depth, morphometry (shape) of the water body, hydrological conditions, sediment distribution, accumulation areas of fine-grained sediments, climatic conditions, etc.
6. Select sampling stations within and outside the study area, for example, for collection of sediments outside the study area at a selected control site. Plot the sampling stations on a chart containing the study area. Number the sampling stations in the most logical sequence relevant to the sediment sampling program and objectives.
7. Select the time frame of the sediment sampling program. Consider the optimal use of the time spent on the sediment sampling in the study area.
8. Consider the safety of the personnel carrying out the sampling program, such as weather conditions expected during the sampling period (wind speed and direction, air and water temperature) and severity of contamination of the sediments and water to which the sampling personnel will be exposed.
9. Select and list all sediment sampling equipment and other materials that will be used in support of the sediment sampling program in the study area, such as tools and spare parts for emergency repairs and maintenance of the sampling equipment in the field; maps; charts; note books; sediment logging sheets; equipment for measuring sediment properties in the field, such as pH and Eh meters, etc.; equipment for homogenization and subsampling of the sediment samples; sample containers; extruders for sediment cores; storage boxes; and other equipment specific to the sediment sampling program.
10. List the last date the sediment sampling equipment was tested together with any problems encountered during the testing and repairs of the equipment. List all necessary spare parts and tools that must accompany the sediment sampling equipment for emergency repairs during the sampling program.
11. Select and compile appropriate sediment sampling and subsampling procedures, sample handling, sample preservation, field storage, transport from the site to the laboratory, and storage after samples delivery including required temperature, freezing of the samples, etc.

2.10 EXAMPLES OF SELECTION OF SAMPLING STATIONS IN STUDIES OF BOTTOM SEDIMENTS

Selection of the location and number of sediment sampling stations in the study area is governed by the study objectives. Examples given below are different methods for the selection of the location and number of sampling stations in sediment sampling programs with various study objectives in lakes, rivers, nearshore freshwater and marine areas, harbors, etc. The selection of sampling stations in the examples uses a statistical and nonstatistical approach in bottom sediment sampling within different water bodies. The objectives of the sediment sampling program are outlined in each example.

The examples below show statistical methods, such as stochastic random (with some modifications), stratified random, and systematic regular, used in the selection of the location and number of the sediment sampling stations. The nonstatistical methods are basically opportunistic and judgmental, i.e., deterministic. Generally, in the stochastic random method, the water body to be sampled is divided into equally sized, numbered subareas, such as blocks or triangles. In summary, random numbers are selected from tables or may be electronically generated. The sediment samples are collected in the subareas chosen by the selected or generated random numbers, usually in the center of each of the chosen subareas. In the stratified random method, the water body to be sampled is divided into a number of blocks, using information on the type of sediments, such as particle size distribution, concentrations of organic matter, etc., or using the water depth. Each of the blocks are divided into subareas that are then randomly selected for sampling, similar to the stochastic random method. In the systematic regular method, the water body to be sampled is divided into a regular grid by a number of blocks. Sediments are sampled in each block at a specific point, such as the center of the block, to keep a constant distance between the sampling stations. The advantage of this method is that the regular sampling grid covers the entire water body to be sampled. However, in a water body with expected great heterogeneity in sediment type and chemistry, such as rivers or small lakes with a complex morphometry, a large number of samples need to be collected. Although it may be not necessary to analyze all collected sediment samples, the funds and time spent on the sediment sampling may be considerable.

Selecting the location and number of sediment sampling stations in a water body using nonstatistical methods is based on available information on the study area, such as historical contamination, sediment distribution, outfalls and other point sources of contaminants entering the water body, etc. In both statistical and nonstatistical methods, the selection of the number of sediment sampling stations depends on the study objectives, size of the study area and water body to be sampled, and economic feasibility, i.e., availability of funds, time, and personnel to carry out the sampling program.

2.10.1 Examples of Selection of Sampling Stations Using a Statistical Approach

Dredging Studies

The study area, which is usually the proposed dredging site, is divided into numbered blocks or triangles. Random numbers, selected from tables or electronically generated, are used to select the sediment sampling stations according to the number of required samples. The number of samples to be collected

Table 2-6 Number of Sediment Blocks to be Sampled Based on the Quanity of Sediments to be Dredged

Volume to be dredged (m^3)			Number of samples
0	to	5 000	3
5 000		6 400	4
6 400		8 200	5
8 200		10 000	6
10 000		16 667	7
16 667		23 333	8
23 333		30 000	9
30 000		36 667	10
36 667		43 333	11
43 333		50 000	12
50 000		58 333	13
58 333		66 667	14
66 667		75 000	15
75 000		83 333	16
83 333		91 667	17
91 667		100 000	18
100 000		140 909	19
140 909		181 818	20
181 818		222 727	21
222 727		263 636	22
263 636		304 545	23
304 545		345 455	24
345 455		386 364	25
386 364		427 273	26
427 273		468 182	27
468 182		509 091	28
509 091		550 000	29
550 000		590 909	30
590 909		631 818	31
631 818		672 727	32
672 727		713 636	33
713 636		754 545	34
754 545		795 455	35
795 455		836 364	36
836 364		877 273	37
877 273		918 182	38
918 182		959 091	39
959 091		1 000 000	40

From Environment Canada, Conservation and Protection, Québec Region (EC, CP, QR), *Sediment Sampling and Preservation Methods for Dredging Projects*, 1987. With permission.

depends on the size of the area to be dredged. The locations of the sampling stations are usually in the center of the selected blocks.

The number of blocks to be sampled recommended by Environment Canada, Conservation and Protection, Québec Region (1987) is based on the project size, i.e., the quantity of sediments to be dredged in cubic meters (Table 2-6). The number of samples to be collected (shown in Table 2-7) is based on work by Atkinson (1985). It is further recommended by EC, CP, QR (1987) that for the proper selection of the sampling stations, the study area should be divided into blocks of equal size, and blocks to be sampled should be chosen using a

Table 2-7 Number of Samples to be Collected as a Function of Project Size

Project size (m^3)	Number of samples to be collected
Very small (<10,000)	3–6[a]
Small (10,001–50,000)	7–12
Average (50,001–100,000)	13–18
Large (100,001–1,000,000)	19–40
Very Large (>1,000,001)	41 + (volume – 1,000,000)/75,000[b]

[a] It is assumed that six samples are sufficient to obtain an acceptable level of confidence for a project of 10,000 m^3 (Ocean Chem Sciences Ltd., 1984).

[b] Atkinson (1985).

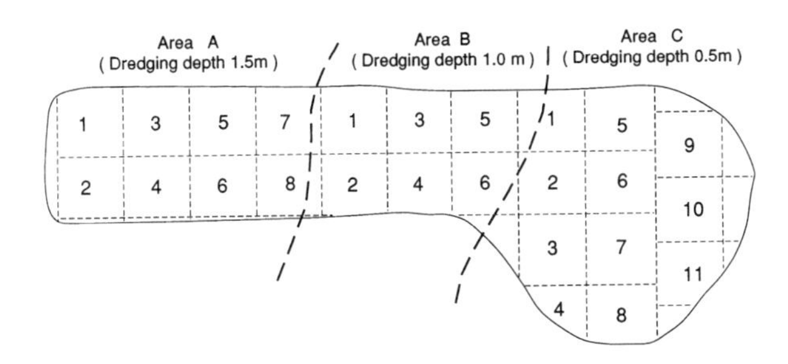

Figure 2-5 Example of selection of sampling stations at an area with different dredging depths (after Environment Canada, Conservation and Protection, Québec Region, 1987).

random number table. For a project in which less than 50,000 m^3 of sediments have to be dredged, the blocks should not exceed 625 m^2 in size, i.e., 25 m × 25 m. For projects where larger quantities of sediments will be dredged, it is recommended to divide the study area into at least five times more blocks than the recommended quantity of samples. Further, the block area should be relevant to the number of the blocks and to the size of the study area. If the dredging depth varies or the study area is irregular, it is recommended to divide the study area into two or more zones. The number of blocks in each zone should be calculated from the total volume of sediment to be dredged, and the number of samples taken in each zone should be proportional to the quantity of the sediment dredged in the zone. An example of this method is given in Figure 2-5. In study areas located in a heavily industrialized watershed, EC, CP, QR recommends more intensive sampling by increasing the number of the sampling stations. In study areas where bottom sediments are expected to have a very heterogeneous chemical composition, such as areas with many industrial and municipal outfalls and other point sources, a stratified sampling design is recommended similar to that described below.

In summary, EC, CP, QR recommends random selection of sampling stations in most cases. The stratified selection should be used in cases where information is needed on contaminant distribution in the sediments in the study area, i.e., spatial variability, contaminant mobility, and the effects of specific sources of contamination. To our knowledge, the sediment sampling method recommended by EC, CP, QR was never tested in the field. However, it is used in this manual as an example of a method that may be suitable in dredging studies because of lack of other suitable methods.

In another example, interim sampling guidelines were developed by Environment Canada to provide information on contaminant concentrations in sediments designated for disposal in the sea under the Canadian Environmental Protection Act (CEPA). The guidelines use a management unit of 1,000 m^3 to describe the division of the project area into a number of blocks to be sampled. In selecting the sampling stations, it is assumed that any location of sediments with contaminant concentrations greater than given in the regulations is randomly distributed within the study area. The following main steps of the guidelines were described by MacKnight (1994).

1. Define the dredging area (identical in this case with the study area) in m^2 or km^2 and volume of sediment in m^3 to be dredged.
2. Calculate the dimensions of a sampling block using 1,000 m^3 as a management unit, and the thickness of sediment planned for removal. For example, if the required thickness of the sediment to be removed is 1 m and the management unit is 1,000 m^2, then the area of the sampling block is 1,000 m^2, which corresponds to a square of about 32 m × 32 m. On the other hand, if the thickness of the sediment to be dredged is 0.2 m, the area of the sampling block is 5,000 m^2, which corresponds to a square of about 71 m × 71 m.
3. Calculate the number of the sampling blocks by dividing the entire project area by the dimensions of one sampling block.
4. If no historical data are available for either fine-grained sediments or contaminants distribution at the study area, the other 60% of blocks are sampled on the basis of truly random selection. Sampling stations are designated at the center of each selected block. An example of this case is outlined in Figure 2-6. According to the bathymetry, a uniform removal of 1 m of sediment is required. Ten management units/blocks are defined for sediment sampling.
5. If a significant outfall or other point source of contamination is known to enter the study area, the block into which the effluent enters must be sampled in addition to sampling the 60% of all blocks. An example of such a case is shown in Figure 2-7. The sampling pattern is adapted to the need to sample the area influenced by the defined point source.

When the distribution of the fine-grained sediment in the study area is known, the blocks must be designated to contain gravel, sand, or fine-grained

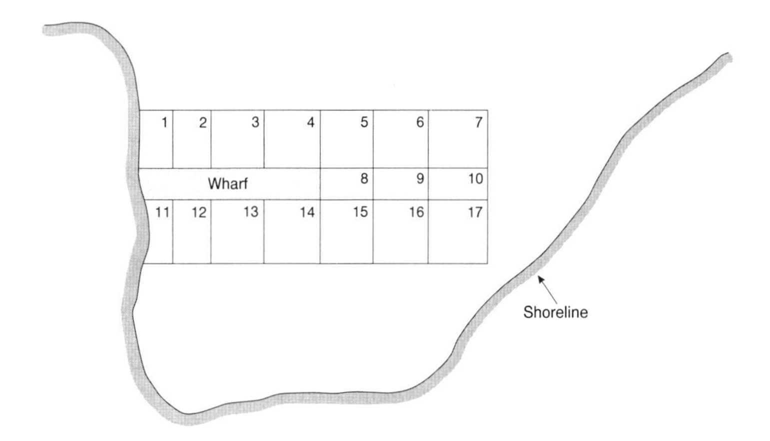

Figure 2-6 Selection of sampling stations at an area for which no data are available (after MacKnight, 1994).

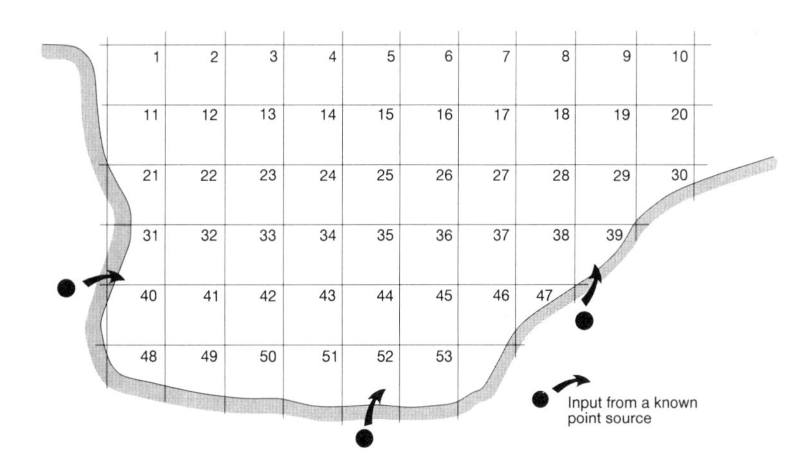

Figure 2-7 Selection of sampling stations at an area with a known point source (after Macknight, 1994).

sediments on the basis of median particle size of the sediments. Again, 60% of the total number of blocks is to be sampled, but the same proportions of gravel:sand:fine-grained sediment must be maintained as for the overall project. These sediment types have to be sampled at a minimum of one sampling station. An example of this case is given in Figure 2-8.

In both methods outlined above, the stochastic random and stratified random, sediment samples have to be taken at each designated sampling station as a core or borehole to ensure a complete vertical description of the material to be dredged. Each core or borehole sample is subdivided into horizontal sections of a specific size (for example, into 5-, 10-, or 15-cm sections). To constrain costs, it is recommended that the topmost section of each core or borehole be analyzed first and the concentrations compared for the contaminants of interest to the regulations. If the concentrations of any one contaminant exceed the regulations, a deeper section of the core/borehole should be analyzed until a section is reached in which contaminant concentrations are lower than stipulated in the regulations. Those sediments containing greater concentrations of contaminants than the regulations allow usually have to be disposed of in a confined disposal facility or treated by various methods after their removal from the bottom. The remainder of the sediments is acceptable for open-water disposal.

If the results of sediment analyses indicate blocks with a high concentration of contaminants, then either the sediment from the remaining 40% of the originally designated blocks should be sampled and analyzed, or the block(s) with contaminated sediments should be subdivided into smaller blocks, i.e., into a secondary pattern. Sediment samples are collected from these blocks in the same manner as for the first pattern. The purpose is to define, as closely as is reasonable, those areas requiring special methods for disposal of the contaminated sediments.

Environment Canada's guidelines emphasize the use of historical data to assist in planning the sampling design. Incomplete or outdated data should be

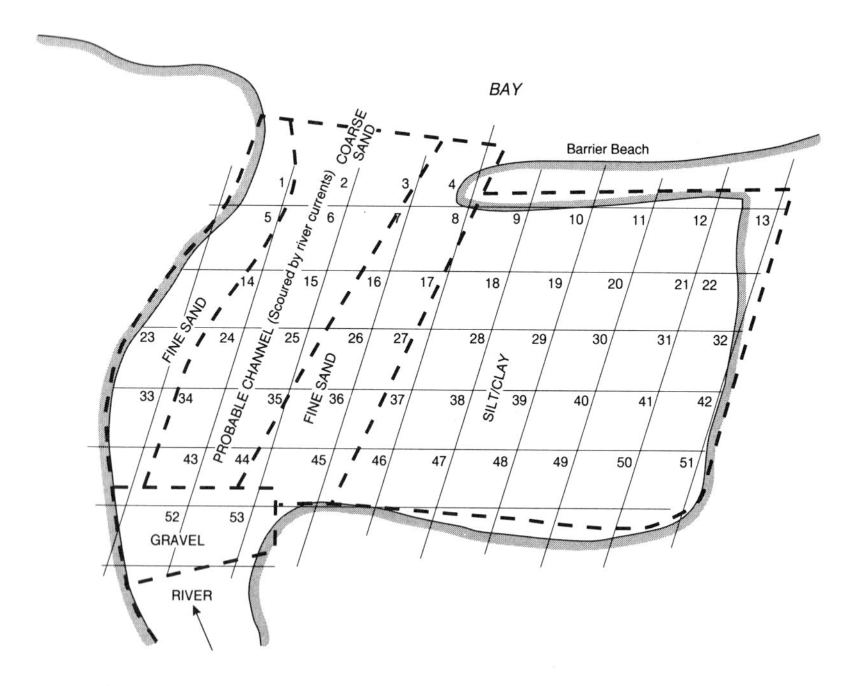

Figure 2-8 Selection of sampling stations at an area with a known particle size distribution in sediments (after MacKnight, 1994).

used carefully, but can still provide useful information in the design of the sampling stations.

Studies of Sediments in Lakes

In the early 1970s, sedimentology and geochemistry of surficial sediments in the Great Lakes of North America were investigated (Thomas et al., 1972; Thomas and Mudroch, 1979). The objectives of this pioneering, large-scale investigation were to obtain information on sediment distribution, geochemistry of Great Lakes sediments, and concentrations of metals in surficial sediments to identify possible sources and inputs of the metals into the Great Lakes. Sediment sampling stations were located at alternate intersections of a 10-km grid based on the Universal Transverse Mercator (UTM) coordinate system (Figure 2-9). In some areas of the Great Lakes — for example, the western basin of Lake Erie — the regular sampling grid was intensified by adding more sampling stations because of expected contamination of the sediments in the area, or sudden changes in bathymetry that could have affected the sediment distribution.

At each sampling station, three Shipek samples were collected. The sediment from one of the Shipek buckets was processed for identification of benthic organisms. Sediment pH and Eh were measured in one of the collected buckets; the top 3 cm of sediment were subsampled from the two Shipek buckets and used for sedimentological and geochemical analysis. The selection of the location of sampling stations using the regular grid approach generated a large number of sediment sampling stations in each of the Great Lakes — for example, 192 and 408 sampling stations in Lakes Huron and Superior, respectively. However, the dense sampling grid and low frequency echosounding between the sampling stations enabled the results of sediment analyses to be used to generate the first detailed maps of sediment distribution (Figure 2-10) and horizontal distribution of metals in surficial sediments in the Great Lakes (Figure 2-11).

2.10.2 Selection of Sampling Stations Using a Non-Statistical Approach

Dredging Studies: Assessment of Sediment Quality in Harbors

The objective of the study was to evaluate the quality of bottom sediments in small-craft harbors located along the Canadian shoreline of Lakes Ontario, Erie, and St. Clair (Thomas and Mudroch, 1979). The harbors are regularly or occasionally dredged to maintain the water depth for use by small boats, such as recreational or small commercial fishing boats. The results of the study were used to make a recommendation on the disposal of the dredged sediments. Sediment sampling stations in the study were selected in the portion of each

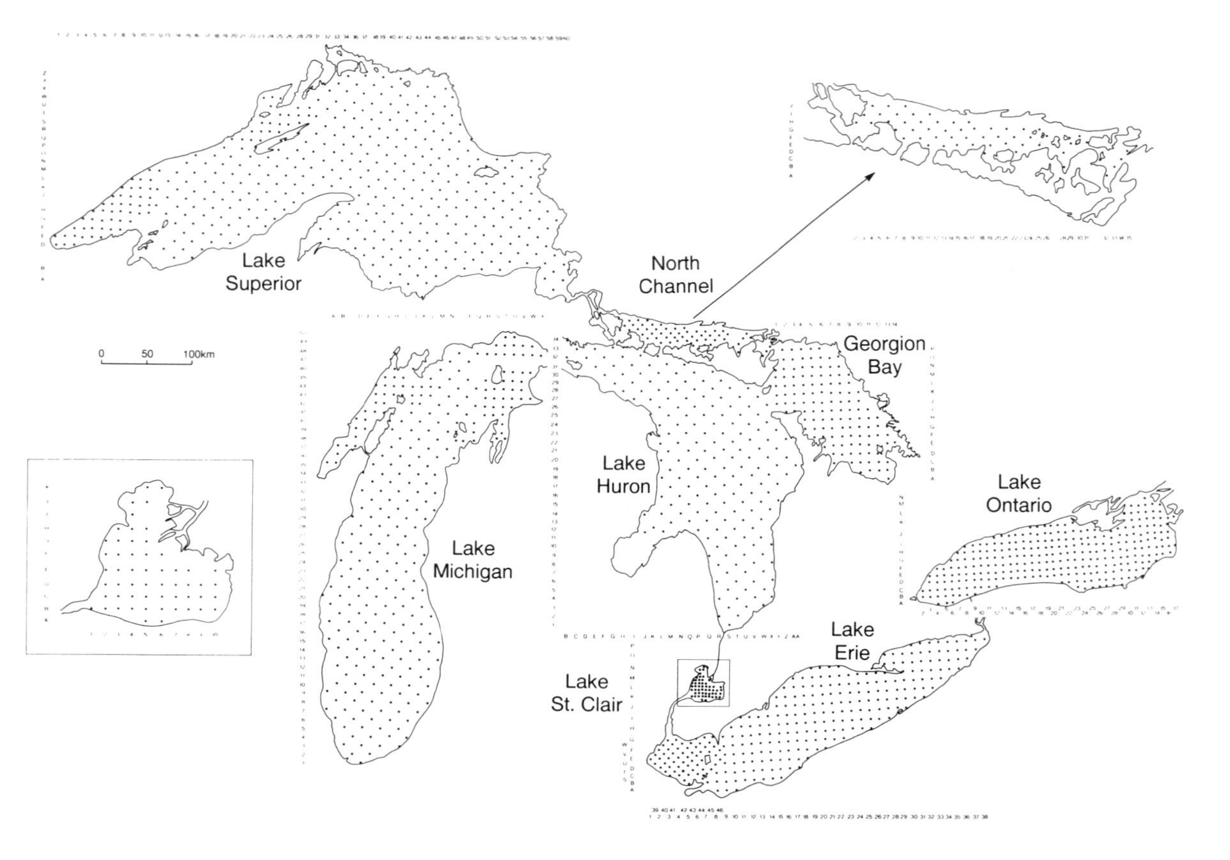

Figure 2-9 Sediment sampling stations in the Great Lakes of North America (after Thomas and Mudroch, 1979).

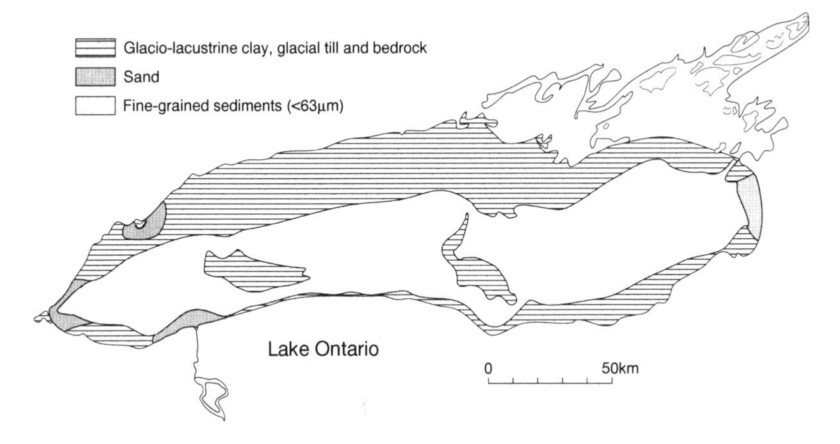

Figure 2-10 Sediment distribution in Lake Ontario (after Thomas et al., 1972).

harbor where the dredging took place. The number of sampling stations was selected according to the size of the harbor and a preliminary survey in each harbor to determine the deposition of fine-grained sediments. The number of the sampling stations was greater within the areas with fine-grained sediments than within those with sand or gravel on the bottom. The preliminary survey was carried out immediately prior to the sediments sampling program, and involved collection of surficial sediments by a small grab sampler, such as mini-Ponar or mini-Shipek, within the area to be dredged. Figure 2-12 outlines the locations of the harbors of Lake Erie and shows the location and number of selected sampling stations within the harbors. In each harbor, one or few sediment cores were collected in addition to the surficial sediments to evaluate vertical sediment quality. For the sediment coring, sampling stations were selected within areas of fine-grained sediment accumulation.

Effects of a Point Source on Sediment Quality

In an area where the effects of an outfall or other point source are expected to affect the sediment quality, the selection of the location and number of sampling stations is based on the expected decrease in impact with increasing distance from the point source. Figure 2-13 outlines a sampling grid along the point source within a water body, such as dispersion of drilling mud residues from an off-shore drilling platform, dispersion of disposed dredged material from a barge, etc. (MacKnight, 1994). The sampling grid would begin with a fixed distance x from the point source; sampling stations would be located at points of $2x$, $4x$, $8x$, etc., toward the area that will most likely be unaffected by the point source. The sampling stations are located at the intersection of each distance line and each major point of the compass, i.e., north, south, west, and east of the point source. The sampling stations can be located in a more frequent pattern (i.e., north, northwest, south, southwest, etc.), as outlined in Figure 2-13. However, it is important to remember that currents and other

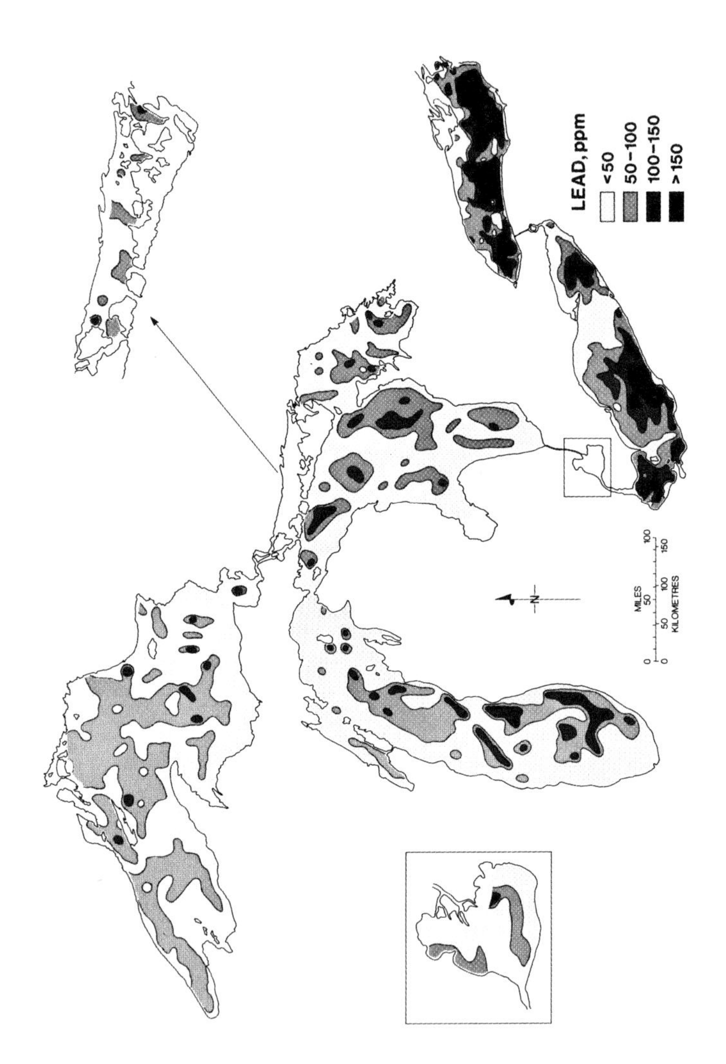

Figure 2-11 Distribution of lead in surficial sediments in the Great Lakes of North America (after Thomas and Mudroch, 1979).

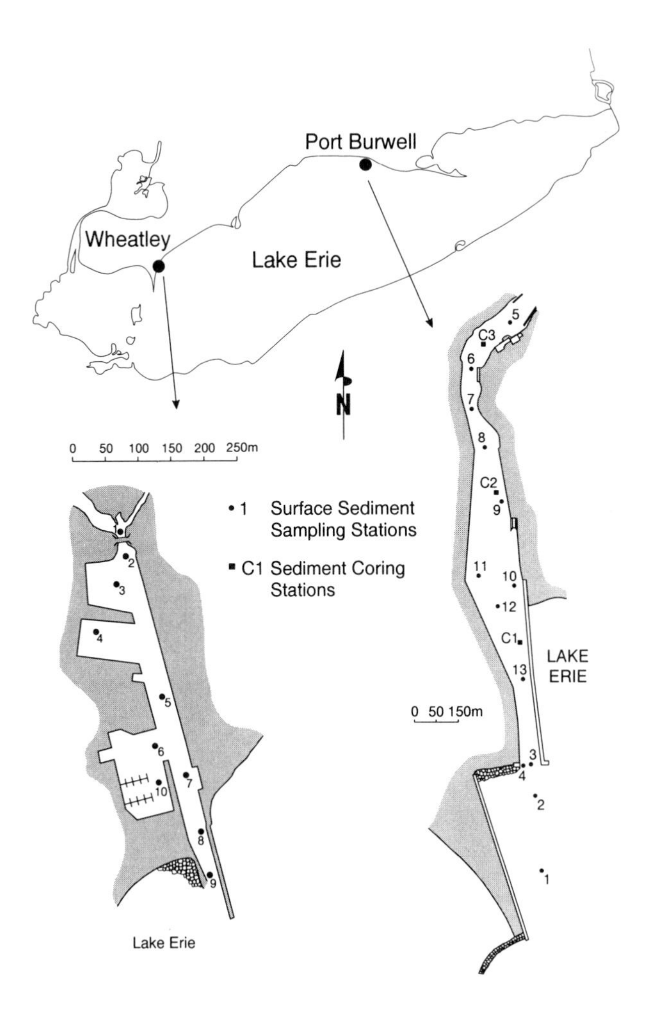

Figure 2-12 Location of sediment sampling stations in harbors on Lake Erie (after Thomas and Mudroch, 1979).

physical factors may affect the dispersion of the contaminants from the point source.

Studies of Bottom Sediments in Rivers, Estuaries, and Bays

Bottom sediments were sampled in ten rivers of the Madeira River watershed, Amazon, Brazil, to characterize the rivers' environment (Lacerda et al., 1990). The objectives of the study were to determine the distribution and geochemical partitioning of trace elements in bottom sediments in white, clear, and black rivers of the Madeira River watershed, to assess the validity of the classical rivers classification in application to sediments. Twenty-eight sampling stations were established in the ten rivers of the Madeira River watershed (Figure 2-14). The sampling stations were located preferentially close to the mouths of the rivers to integrate the major biogeochemical processes occurring in each subbasin. Obtained sediment samples were oven-dried at 80°C in the laboratory to a constant weight, and silt and clay particles (<63 μm) were separated by sieving and analyzed to determine their chemical composition. The location of the 28 sampling stations and the analyses of the collected sediments enabled the authors of the study to conclude that the classical classification of Amazon rivers based on hydrochemical parameters can also be applied to bottom sediments for most of the determined elements.

Bottom sediments were collected at seven and five sampling stations in Göksu River and Taçucu Delta, respectively, Turkey (Figure 2-15) to determine the chemical composition of the sediments (Sanin et al., 1992). The objectives of the study were to assess the transport of sediment-associated metals by the river and the influence of the river on trace metal concentrations in bottom sediments of the adjacent part of the Mediterranean Sea. Twenty-six surface sediment samples were collected at seven sampling stations thought to

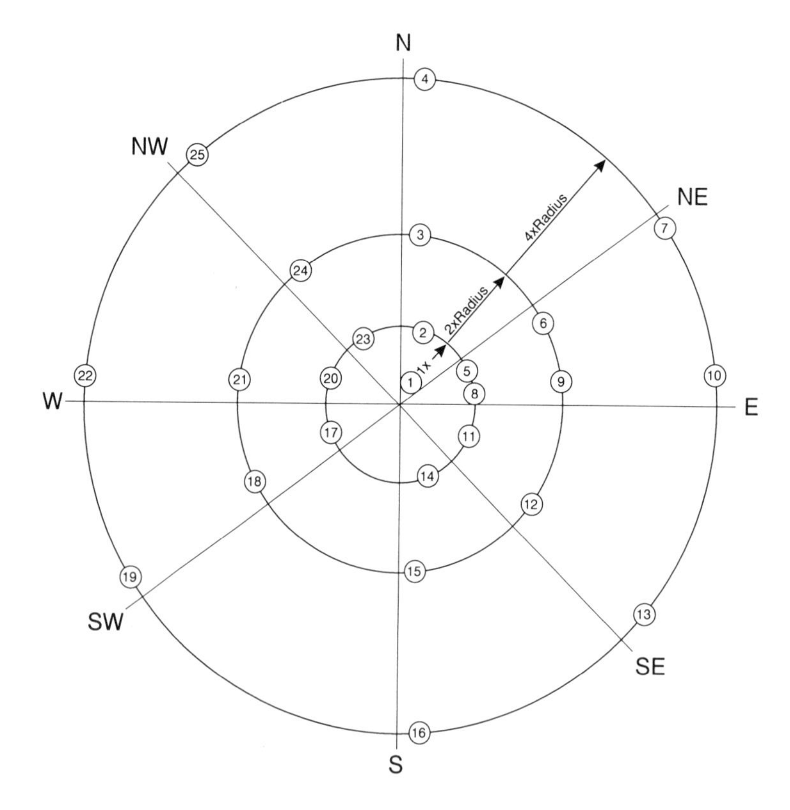

Figure 2-13 Selection of sampling stations along a known point source (after MacKnight, 1994).

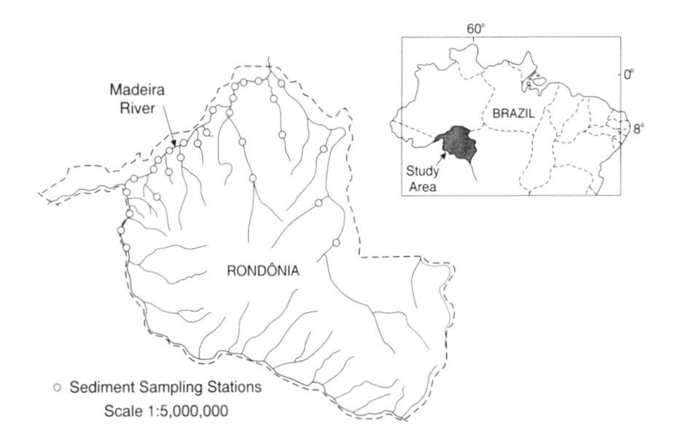

Figure 2-14 Selection of sampling stations in the Madeira River and its tributaries, Brazil (after Lacerda et al., 1990).

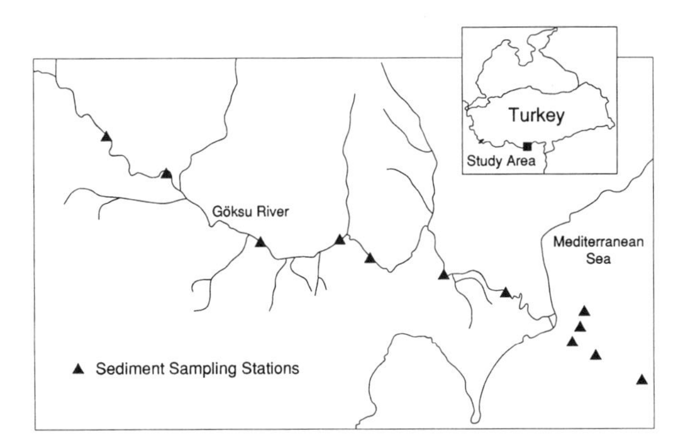

Figure 2-15 Selection of sampling stations in Göksu River and along the estuary, Turkey (after Sanin et al., 1992).

be representative of the river body and its Mediterranean delta. The freshwater sites were located up to 100 km upstream from the river mouth. Marine sediment samples were collected by a Van Veen-type grab sampler from five sampling stations located at three water depths: 10 m, 20 m, and 50 m. Samples were collected from the same embankment throughout the river by coring. Collected sediments were used in the quantitative determination of different major and trace elements. The authors of the study concluded that the obtained data set may be used as a baseline for comparing environmental factors in other rivers.

Bottom sediments were collected in a study of sediment quality in Cardiff Bay, U.K. (Hitchcock and Thomas, 1992). The objectives of the study were to determine the quality of the sediments and trace metals distribution in the Cardiff Bay prior to an action for improvement of the appearance of the bay area and lower reaches of its tributaries, Rivers Taff and Ely. The sediment samples were obtained at 18 sampling stations in Cardiff Bay (Figure 2-16). The location of the sampling stations was selected to correspond to the major stream axes of the Rivers Taff and Ely. The location was also related to the proposed dredging operations in the bay. Data obtained from physico-chemical analyses of the sediment samples were considered a part of baseline data for the activities for improvement of the Cardiff Bay area.

Heterogeneity of spatial distribution of metals in sediments has been studied in Monvallina Bay, Lake Magiore (Baudo and Muntau, 1990). Extremely dense sampling stations in a grid (i.e., 30 sampling stations per km^2) were used in the study (Figure 2-17). Lake-bottom morphology affected sediment sorting and deposition, resulting in complex distribution patterns of metals of interest. Data obtained by the analysis of collected sediments were used in mapping the distribution of different metals in the bay. The distribution of vanadium shown in Figure 2-17 indicated possible sources of the element. The maps can be further used to assess the origin of the metal and its pathways and fate, and as background information for further studies, such as of the release of the sediment-associated metal and its effects on biota. Different aspects of mapping the distribution of contaminants in sediments and methods used in the mapping were described by Baudo (1990). Figure 2-18 shows an example given by Baudo on the distribution of sediment sampling stations in Lake Orta and a map of the distribution of copper in the lake bottom sediments. The map shows areas of different concentrations of copper in sediments that were

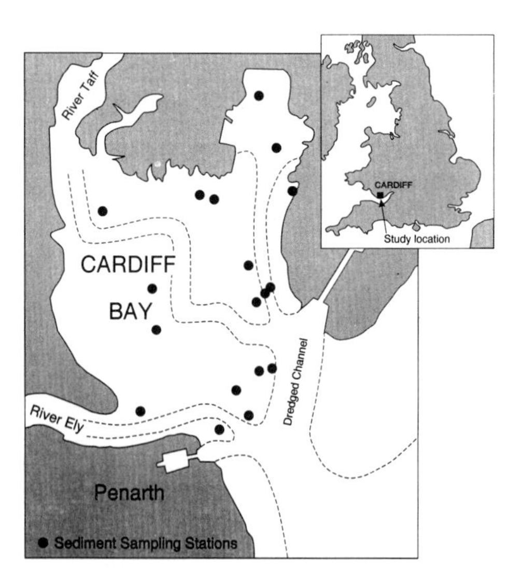

Figure 2-16 Selection of sampling stations in Cardiff Bay, U.K. (after Hitchock and Thomas, 1992).

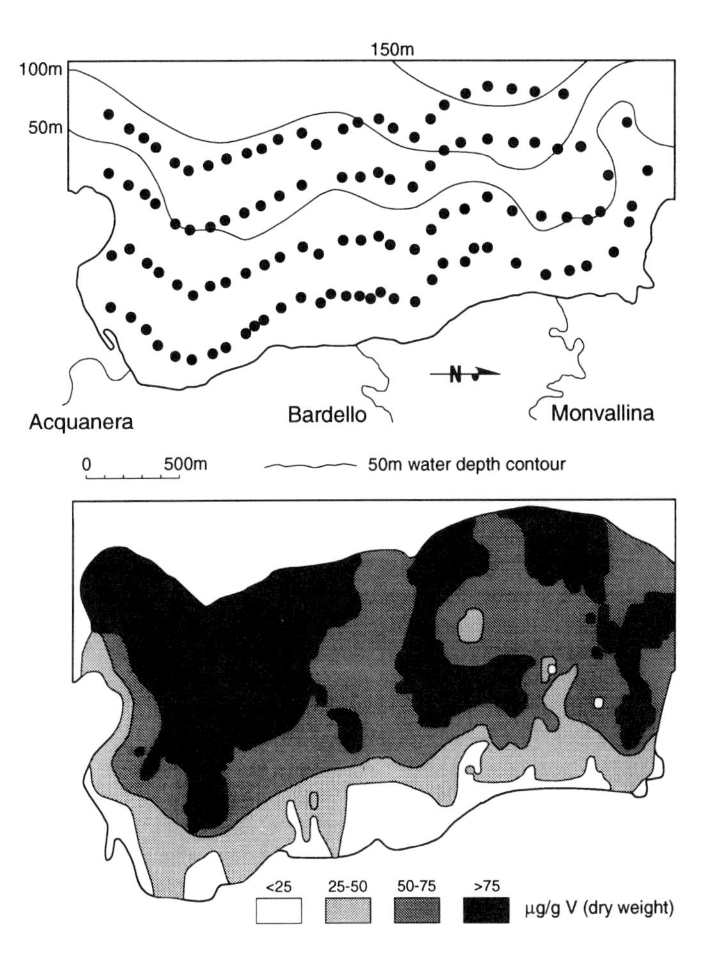

Figure 2-17 Sampling stations grid and mapping of concentration of vanadium in Monvallina Bay, Lake Magiore, Italy (after Baudo and Muntau, 1990).

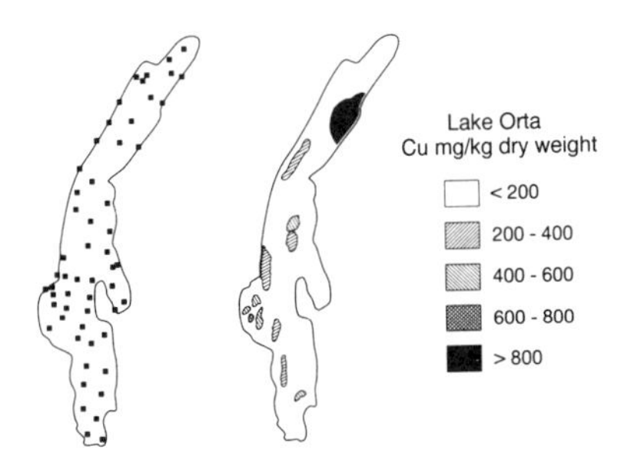

Figure 2-18 Location of sampling stations and the distribution of copper in surficial sediments in Lake Orta, Italy (after Baudo, 1990).

determined by six different mapping techniques. From the results of the mapping, Baudo concluded that since the mapping of the studied variable (i.e., contaminant) is likely to be used for practical purposes — such as the estimation of the whole lake burden or sediment treatment by dredging, capping, or other remediation techniques — at least two different cartographic techniques should be employed, one of the trend-surface type and the other of the moving-average type.

REFERENCES

Ackermann, F., A procedure for correcting the grain size effect in heavy metal analysis of estuarine and coastal sediments, *Environ. Technol. Lett.*, 1, 518, 1980.

Allan, R.J. and Ball, A.J., *An Overview of Toxic Contaminants in Water and Sediments of the Great Lakes*, Monograph Series of Water Pollution Research Journal of Canada, 1990, 680.

Atkinson, G., *ODCA Sampling, Project U 390*, Environment Canada, Environmental Protection Service, Ocean Dumping Program, Internal Document, 1985, 17.

Baudo, R., Sediment sampling, mapping and data analysis, in *Sediments: Chemistry and Toxicity of In-Place Pollutants*, Baudo, R., Giesy, J., and Muntau, H., Eds., Lewis Publishers, Ann Arbor, 1990, 15.

Baudo, R. and Muntau, H., Lesser known in-place pollutants and diffuse source problems, in *Sediments: Chemistry and Toxicity of In-Place Pollutants*, Baudo, R., Giesy, J., and Muntau, H., Eds., Lewis Publishers, Ann Arbor, 1990, 1.

Crépin, J. and Johnson, R.L., Soil sampling for environmental assessment, in *Soil Sampling and Methods of Analysis*, Carter, M.R., Ed., Lewis Publishers, Boca Raton, 1993, 5.

de Groot, A.J., Zschuppe, K.H., and Salomons, W., Standardization of methods of analysis for heavy metals in sediments, *Hydrobiologia*, 92, 689, 1982.

Dowring, J.A. and Rath, L.C., Spatial patchiness in the lacustrine sedimentary environment, *Limnol. Oceanogr.*, 33, 447, 1988.

Environment Canada, Conservation and Protection, Québec Region (EC, CP, QR), *Sediment Sampling and Preservation Methods for Dredging Projects*, EC, CP, QR, Québec, 1987.

Förstner, U., Accumulative phases for heavy metals in limnic sediments, *Hydrobiologia*, 91, 269, 1982.

Håkanson, L., The influence of wind, fetch, and water depth on the distribution of sediments in Lake Vanern, Sweden, *Can. J. Earth Sci.*, 14, 397, 1977.

Håkanson, L. and Jansson, M., *Principles of Lake Sedimentology*, Springer-Verlag, Berlin, 1983, 32.

Hitchcock D.R. and Thomas, B.R., Some trace metals in sediments from Cardiff Bay, U.K., *Mar. Pollut. Bull.*, 24, 464, 1992.

Horowitz, A.J. and Elrick, K.A., Interpretation of bed sediment trace metal data: methods for dealing with the grain size effect, in *Chemical and Biological Characterization of Sludges, Sediments, Dredge Spoils, and Drilling Muds*, ASTM STP 976, Lichtenberg, J.J., Winter, J.A., Weber, C.I., and Fradkin, L., Eds., American Society for Testing and Materials, Philadelphia, 1988, 114.

Keith, L.H., *Principles of Environmental Sampling*, ACS Professional Reference Book, American Chemical Society, 1988, 458.

Lacerda, L.D., De Paula, F.C.F., Ovalle, A.R.C., Pfeiffer, W.C., and Malm, O., Trace metals in fluvial sediments of the Madeira River watershed, Amazon, Brazil, *Sci. Total Environ.*, 97/98, 525, 1990.

Loring, D.H. and Rantala, R.T.T., Manual for the geochemical analyses of marine sediments and suspended particulate matter, *Earth Sci. Rev.*, 32, 235, 1992.

MacKnight, S.D., Selection of bottom sediment sampling stations, in *Handbook of Techniques for Aquatic Sediment Sampling*, 2nd ed., Mudroch, A. and MacKnight, S.D., Eds., Lewis Publishers, Chelsea, Michigan, 1994.

Mudroch A., Particle size effects on concentration of metals in Lake Erie bottom sediments, *Water Pollut. Res. J. Can.*, 19, 27, 1984.

Mudroch, A., Lake Ontario sediments in monitoring pollution, *Environ. Monitor. Assess.,* 28, 117, 1993.

Mudroch, P., Assessment of sediment quality at the Cornwall area of the St. Lawrence River, 1991, Report, Environmental Protection, Ontario Region, Toronto, 1993, 40.

Ocean Chem Sciences Ltd., Background document on sediment sampling strategies and methodologies for dredged marine sediments (preliminary report), Environment Canada, Environmental Protection Service, Ottawa, 1984, 110.

Radio Technical Commission for Maritime Services, *RTCM Recommended Standards for Differential NAVSTAR GPS Services*, Radio Technical Commission for Maritime Services, Washington, 1990, 102.

Sanin, S., Tuncel, G., Gaines, A.F., and Balkas T.I., Concentrations and distributions of some major and minor elements in the sediments of the River Göksu and Taçucu Delta, Turkey, *Mar. Pollut. Bull.*, 24, 167, 1992.

Sly, P.G. and Thomas, R.L, Review of geological research as it relates to an understanding of Great Lakes limnology, *J. Fish Res. Board Can.*, 31, 795, 1974.

Thomas, R.L. and Mudroch, A., *Small Craft Harbours-Sediment Survey, Lakes Ontario, Erie and St. Clair, 1978*, Dredging Summary and Protocol, Report to Small Craft Harbours, Ontario Region from the Great Lakes Biolimnology Laboratory, 1979, 149.

Thomas, R.L., Kemp, A.L.W., and Lewis C.M.F., Distribution, composition and characteristics of the surficial sediments of Lake Ontario, *J. Sed. Petrol.*, 42, 66, 1972.

CHAPTER 3

Description of Equipment for Bottom Sediment Sampling

3.1 INTRODUCTION

The purpose of sampling is to collect a representative, undisturbed sample of the sediment to be investigated. There are many factors that need to be considered in the selection of suitable equipment for bottom sediment sampling. These factors include the sampling plan; the type of available sampling platform (vessel, ice, etc.), location of and access to the sampling site, physical character of the sediments, the number of sites to be sampled, weather, number and experience of personnel who will carry out the sampling, and the budget. Because of these many factors, the standardization of the sampling techniques is difficult. Generally, the selected sampling equipment should recover an undisturbed sediment sample.

Several excellent, comprehensive reviews are available on bottom sediment sampling devices. These reviews have tended to discuss the limitations of the equipment for particular purposes: for example, grab samplers and corers suitable for sampling of benthic organisms (Holme, 1964); a wide variety of bottom sediment samplers designed for biological and geological work mainly in the marine environment, but that may be used in pollution studies in the marine or freshwater environments (Hopkins, 1964); sediment sampling techniques in studies of sedimentary structure (Bouma, 1969); description and evaluation of performance of commercially available and custom-made bottom sediment samplers (Kajak, 1969); review and testing of bottom sediment samplers used in the lacustrine environment (Sly, 1969); description of different instruments, including bottom sediment samplers, and their use in oceanographic research (Martinais, 1971); efficiency of grab samplers and corers in benthic organisms sampling (McIntyre, 1971a); description of bottom sediment samplers suitable for studies of sediment microbiology (Collins, 1977); bibliography of samplers for benthic invertebrates (Elliot and Tullett, 1978, 1983); bottom sediment sampling strategies and sampling devices in marine studies (Moore and Heath, 1978); sampling devices, including bottom sediment samplers, for studies of marine pollution (Bascom, 1979); description and evaluation of different bottom samplers for studies of geology of the continental shelf (Fowler and Kingston, 1979; Le Tirant, 1979a,b); equipment currently used and under development to investigate the ocean bottom for surveying offshore construction sites (Sly, 1981); and theoretical and practical aspects and advantages and disadvantages of various types of sediment samplers (Håkanson and Jansson, 1983).

Many samplers described in the literature are only variations of a few early models modified to overcome observed deficiencies or to be used for specific objectives and for various operating conditions. The many different names applied to sediment samplers are often confusing to those who must choose one suitable for a project.

In addition to the above reviews, commercially available or custom-made sediment samplers have been described in studies involving bottom sediments in marine and freshwater systems, for example, investigations of benthic fauna, collection of samples at sites with different water depths and sediment textures for geological interpretation, or performance comparisons of newly designed samplers for a specific task.

3.2 CHOICE OF EQUIPMENT FOR SEDIMENT SAMPLING

There are two primary sediment zones of interest in contaminant studies: the surficial or upper 10 to 15 cm, and the deeper layers. Sampling of the surface layer provides information on the horizontal distribution of parameters or properties of interest for the most recently deposited material, such as particle size distribution or geochemical composition of the sediments. The information obtained from analysis of surface sediments can be used, for example, in mapping the distribution of the element of interest in sediments across a lake (Figure 3-1). A sediment column, which includes the surface sediment layer (10 to 15 cm) and the sediment underneath this layer, is collected to study historical changes in parameters of interest or to define zones of pollution. The information obtained by collecting cross sections of sediment by coring can be used in plotting concentration profiles of element of interest (Figure 3-2). The "typical" geochemical profile of a contaminant in a sediment core shows an exponential decrease of contaminant concentrations with sediment depth to a "background" concentration, since many chemical compounds of environmental concern are of recent origin. Therefore, the choice of the type of sediment samples to be collected (i.e., surface sediments or cores) depends entirely on the objective of the study.

The quality of sediments to be collected at each sampling station is another important factor to be considered before choosing the sampling equipment. The required quantity of sediments depends on the number and type of physico-chemical analyses and biological tests that will be carried out. Laboratories

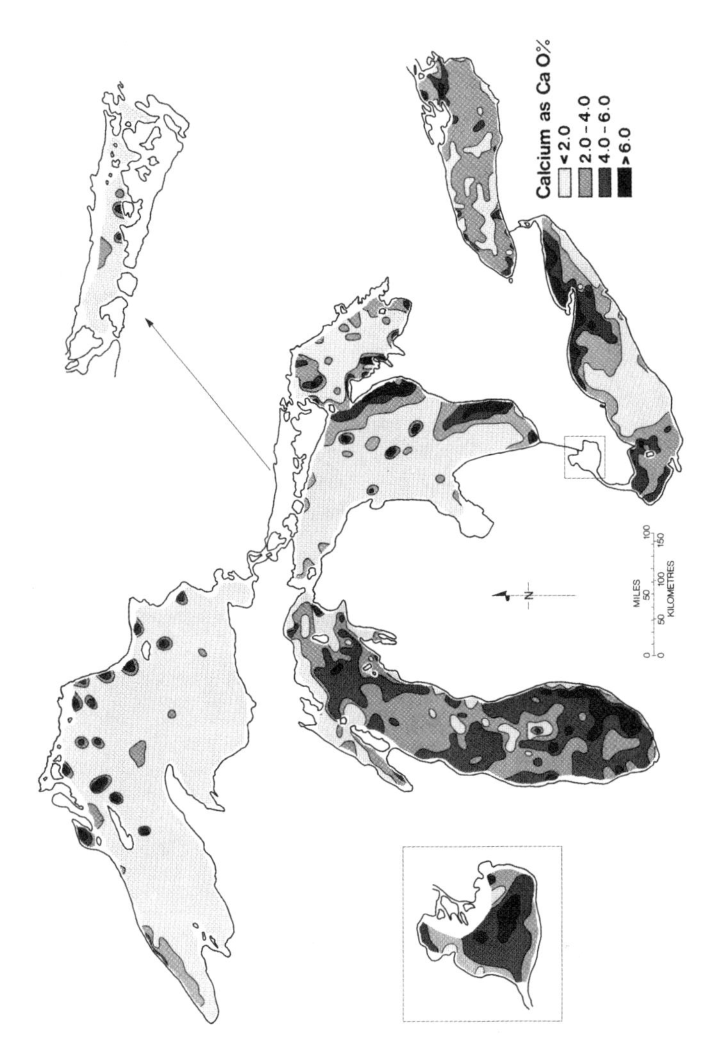

Figure 3-1 Distribution of calcium in surficial sediments in the Great Lakes of North America.

responsible for the analysis and tests should be consulted before sampling to estimate the quality of the sediments. A list of analyses and tests should be prepared showing the quantity of sediments for individual analysis and tests. Further, the laboratory should identify a container for collecting the sample for individual analysis or tests and specify how the sample should be handled or stored and what the sample conditions should be for the analysis (i.e., wet, oven-dried, freeze-dried, etc.). This information should be included in the sediment sampling protocol as discussed in Chapter 2, and is necessary not only to decide the type of sampler but also to consider the number of replicate samples per sampling station, compositing the replicates, etc. An additional quantity of sediments for QA/QC and banking of the samples for future studies also needs to be considered. The containers and conditions to preserve sediment samples for different types of analysis are discussed in Chapter 6.

Other factors that should be considered in the selection of sediment sampling equipment are outlined in Figure 3-3. Access to the sampling area plays an important role in the sampling strategy and logistics and selection of the sampling equipment. There are basically two options for collecting bottom sediment samples: sampling from a platform, and sampling by a diver. A sampling platform could be a vessel, ice, a plane, or a helicopter. Collection by

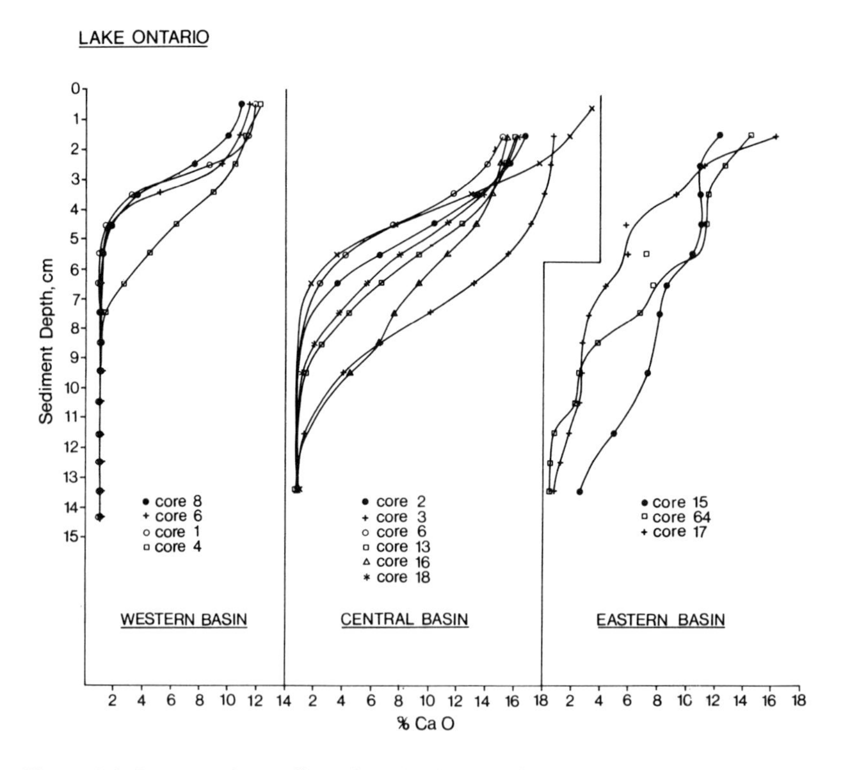

Figure 3-2 Concentration profiles of calcium in Lake Ontario sediments.

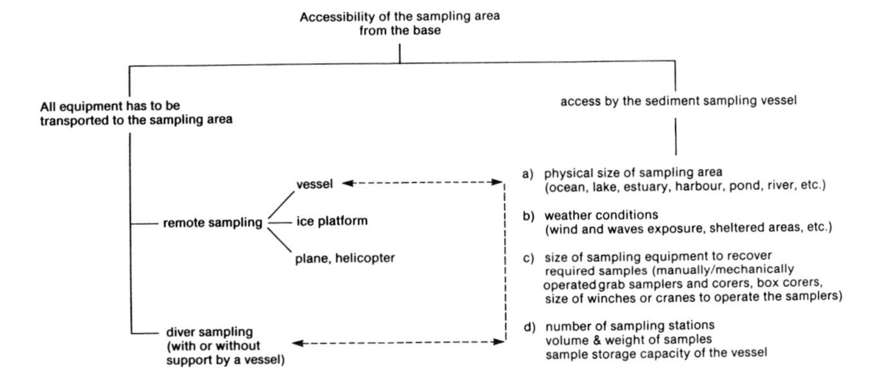

Figure 3-3 Factors to be considered in selection of sediment sampling equipment.

a diver, usually more costly and difficult than sampling from a platform, often yields better-quality samples, particularly sediment cores. In areas with sufficient ice cover over the sampled water body, usually during the winter, sediment samples can be obtained by drilling a hole in the ice and sampling through this. In study areas with no road access, sediments may be collected from a small float plane or from a helicopter. Availability of a suitable plane or a helicopter and cost are factors to be considered. The different options are discussed in more detail below.

3.2.1 Factors to be Considered in the Selection of Bottom Sediment Samplers

Bottom Sediment Sampling from a Vessel

The two primary factors in vessel suitability are the size and depth of the water body being sampled and the proposed method of sampling: surface sediments or sediment cores. The vessel should offer reasonable on-board working space, particularly for subsampling collected sediments and for measurements that need to be made immediately after sample retrieval. Large and heavy coring equipment dictates the use of a vessel with sufficient deck space and lifting capacity for sampling operations. Alternatively, the availability of a specific vessel will affect the selection of sampling equipment that can be safely operated from that vessel. Small grab samplers and corers can be hand-operated from a small vessel. Light, portable winches, hand reels, or line keepers are commercially sold, usually by companies supplying sediment sampling equipment. When more sediment samples must be collected in an area with a water depth greater than 10 m, portable winches can be easily mounted on the vessel for use with a variety of samplers within weight limitations.

Large sampling devices usually weigh between 50 and 400 kg empty. When filled with wet sediment they can weigh 125 to 500 kg (and more), and will require extra winch power to counteract the suction effect of cohesive bottom sediments. For example, a winch or crane with a 2,000- to 4,000-kg capacity is required for a box corer. These samplers need a suitable winch or crane to operate with a lift height of 3 to 5 m above the gunnel of the vessel.

Sampler lines, cables, and depth meters are usually available on large vessels, and are commercially sold by different companies. The strength capacity of the lines and cables should be checked regarding combined weights of the sampling equipment to be used and that of the collected sediment.

A sufficient open work area on the vessel's deck and a various number of operators are required for launching and recovering sediment samplers, particularly the larger ones. If samples are to be partially or completely processed onboard, laboratory space with running water, a distilled water supply, and electric power should be included in the vessel equipment. Special equipment, such as a glove box with an inert gas supply necessary for sediment subsampling under anoxic conditions, may also be required. A noncontaminated area for all field work is extremely important in studies of concentrations of trace contaminants in sediments. Lubricants, gasoline, paints, metal parts, and other materials commonly used on vessels can severely contaminate sediment samples and should be avoided in areas designated for handling sediment samples. Another important factor is the storage facility on the vessel. The size, volume, and weight of sediment samples must be considered together with the required temperature at which samples need to be stored. Sediment samples collected for the determination of contaminants should be stored frozen or at 4°C. The need for freezers or refrigerators places an extra burden on available space and electricity. All equipment on the vessel to support a bottom sediment sampling program should be inspected before departure for the field work.

Great attention should be paid to maneuvering the sampling vessel in shallow water. Sly (1969) noted the disturbance of grab samples extending to 0.3 or 0.6 cm depth of the loose surface material (soft, silty clays) by maneuvering a 19.5-m long sampling vessel with a draft of 1.9 m at 5.5 m water depth. In water depths of 4.5 and 3 m, the top 1.9 and 5 cm, respectively, of the sediment was eroded. Generally, the shallower the water depth at the site, the more likely the approach and maneuvering of the sampling vessel will disturb the surface layer of the sediment.

Bottom Sediment Sampling from Ice

The advantages of sampling through the ice rather than from a floating platform include a steady platform from which to operate and a large space at the sampling station for assembly of the equipment and sample handling. There are many serious problems with collecting samples from the ice, such as malfunctioning of sampling equipment at low temperatures, the transport of equipment from one sampling station to the other by a snowmobile or sleigh, poor weather conditions in the winter, and difficult working conditions, particularly in remote areas. Moreover, sampling from the ice can be considered only when there is sufficient thickness of ice and the ice is relatively stable. Information on ice thickness at the sampling area and the area that will be used for equipment transport needs to be obtained before planning the mode of transport for the equipment to the site and to estimate the time for hole drilling. This information can usually be obtained from provincial or federal government agencies. There is no standard procedure or equipment for sampling from the ice. Portability of equipment is the prime factor in equipment selection. Equipment has to be transported to the site, often under difficult and cold operating conditions. Helicopters and small planes can provide transportation of larger equipment and personnel to remote areas. Snowmobiles are suitable for transport over shorter distances, up to 20 km.

Other factors which need to be considered include drilling holes in the ice; positioning sampling stations; shelter and safety requirements; equipment (all parts of a sediment sampler including lines, winch and retrieval equipment, power source, and sample containers); and all equipment for sample handling at the site.

Hole Cutting

Sampling through the ice requires drilling or sawing a hole in the ice at each sampling station. The size of the hole will depend on the equipment chosen for sediment sampling. Small corers without stabilizer fins or small grabs will require a hole of approximately 25 to 30 cm in cross section. Large grabs and corers will require a hole of about 80 to 100 cm cross section. Some devices have been specially designed for sampling from ice. Marine geological investigations in areas of permanently frozen sea involve lowering sampling equipment through natural or artificial holes in the ice cover. Practical limits on auger hole diameters prevent the use of piston corers having conventional tripping arms. Marlowe (1967) described a device allowing collection of piston cores through a hole with a diameter as small as 21 cm. King and Everitt (1980) described a sediment and water sampler designed for use through surface ice for Antarctic conditions. The sampler was based on mechanical suction and could be deployed through a hole about 11 cm in diameter, which could be drilled easily in thick ice with a Sipre ice auger (Journal of Glaciology, 1958).

Commercially available ice augers and power drives are suitable for drilling holes. Gasoline-powered engines or electric motors can be used for driving ice augers. Usually, hand augers can drill about 35-cm diameter holes in thick ice. Drilling up to 45-cm diameter holes requires a gasoline or electric-power auger. Drilling larger holes in ice requires additional power for large-diameter augers. Chain saws are necessary for cutting large holes at a site where more sediment samples have to be collected. The use of an auger or chain saw takes a considerable amount of time. Sufficient time (and energy) should be allowed in the program for ice cutting. Hot water drills have been developed for more rapid drilling of holes in ice for studies in the Arctic (Verrall and Baade, 1982). The state-of-the-art techniques for making access holes through first-year and multi-year Arctic sea ice were reviewed by Mellor (1986). The review includes methods for penetrating the ice, cleaning debris, maintaining the access holes in an open condition, and particularly, gaining access through the ice to ensure that the surrounding ice remains competent for the support of operating equipment and personnel.

Sampler Deployment

In shallow water, lightweight samplers, described for use from small vessels, can be lowered by hand to the bottom through a hole in the ice. Typically, however, a hand-winch or gasoline-powered winch, mounted on a portable tripod stand, is used for sediment sampling at stations with deeper water or when the sampler is particularly heavy. A larger hole or a few holes in the ice are necessary for collecting more undisturbed samples at one site. A small piece of ice floating in the hole, may cause a premature triggering of the corer. Sly and Gardener (1970) described a specially designed portable winch and frame built and tested for use in through-ice sampling programs. The complete system, made in modular form, is easily handled by one or two operators at temperatures as low as –40°C, and was transported by a ski-plane or snow/ice vehicle. This system was suitable for use with a slightly modified vibratory corer.

Portable gasoline-powered engine drives, power generators, and electric drills are heavy and need to be transported to the sampling site by a snowmobile with a sled, a truck, or an aircraft. A heated shelter should be considered when planning to spend a longer time at the sampling site. A portable insulated shelter, a tent, or a transport vehicle can be used as a shelter. At sites where only small samplers will be used, sampling can be carried out in an insulated shelter or a tent built on the ice.

Sampling from the ice demands experience and a strong focus on worker safety and health. First aid and communication equipment, survival gear, food, fuel, etc., must be provided to a party carrying out sampling in remote areas. All sampling equipment should be thoroughly inspected prior to departure to the sampling site.

Figures 3-4 through 3-8 show different equipment used in sediment sampling from ice.

Collection of Bottom Sediment Samples by Diving

The collection of sediment samples by a diver should be considered when undisturbed samples are required, particularly for studies of the sediment-water interface. Moreover, the diver can select a suitable area on the bottom at the sampling site and make notes or take photographs/underwater videos of the bottom and control the operation of the sampler. There are limitations due to water depth, visibility, or currents. The diver's visibility can be obscured if fine-grained sediments are disturbed or the water is turbid.

Figure 3-4 Drilling a hole with an ice auger.

Figure 3-5 Hole cutting with a hand saw.

Figure 3-6 Hole cutting with a chain saw.

Figure 3-7 Removing cut ice blocks.

Figure 3-8 Portable frame for lowering the sampling equipment through the hole in the ice.

McIntyre (1971b) compared sediment samples collected by gravity corers and scuba divers in studies of meiobenthos. His data, which represented the range of variation found throughout the year, showed that cores collected by the diver consistently gave substantially higher counts than the samples collected by gravity corers. This was particularly true for copepods, which were largely restricted to the top 1 cm of mud. The results indicated that the downwash, caused by gravity corers during their descent, dispersed a considerable proportion of the superficial sediments and, therefore, these corers failed to collect all the animals associated with the uppermost sediment layer. This deficiency in gravity corers or any other instrument dropped, even gently, onto the sea or lake bed should be considered in selecting sampling techniques in studies of recently deposited organic matter or pollutants concentrated at the sediment-water interface.

Simple corers, core tubes, or boxes that can be sealed at both ends are suitable for surface sediment sampling and coring by a diver. Several corers of this type were described by Hopkins (1964). The penetration of 30 to 120 cm by a diver-operated core can be achieved without assistance from the surface. Several other samplers were described and successfully used for sediment collection by a diver. The Birge-Ekman box corer was modified for use by scuba divers or a deep submergence research vessel, allowing the precision of sampling to be controlled (Rowe and Clifford, 1973). Other methods and samplers used in sediment sampling by divers were described by Mudroch and MacKnight (1994).

3.2.2 Sampler Transport and Assembly at Sampling Site

The transport of sediment sampling equipment to a remote area is another factor to be considered in the selection of a sampler. The weight and volume of the sampler that needs to be shipped by air usually limits the choice of equipment. Typically, the sampler has to be dismantled into several parts prior to shipping. Unless the person who assembles the sampler upon its arrival at the sampling location is familiar with the equipment, a detailed description, with a simple drawing of the assembling procedure and a list of all parts, should accompany the sampler. A set of tools essential for the assembly should be included in the shipment. Shipping of critical spare parts with the sampler, particularly those that may become lost or damaged during sampling such as springs and pins, often saves time and money.

3.2.3 Sediment Type

There is no one bottom sediment sampler that can be used for the collection of all sediment types. There are many samplers for collecting surface sediments and sediment columns that will recover an undisturbed sample only in soft, fine-grained sediments. Fewer grabs and corers are available for collecting sediments containing sand, gravel, or firm glaciolacustrine clay or till. It is difficult to choose a proper bottom sediment sampler without knowing the bathymetry and areal distribution of physically different sediment types at the sampling site. Consequently, gathering all reported information on the bathymetry and distribution of physically different sediments should be considered for any area to be sampled (see Chapter 2 for further information on sediment type). Consultation with personnel experienced in sedimentological studies will help in the final choice of sampling equipment.

3.2.4 Sediment Depth to be Sampled

The sediment sampling plan should identify the sampling stations and the sediment depth that needs to be sampled at individual locations. In the evaluation of such sediments, there may be regulatory requirements for handling or treatment of sediments in excess of a specific concentration of a contaminant in a manner different from the "uncontaminated" underlying sediments, where this handling or treatment entails considerable cost per cubic meter of material. Detailed characterization can closely define the contaminated sediment horizon, thereby limiting such costs. For example, 1 m of sediment needs to be removed by dredging for navigational purposes. Detailed sampling shows that the contamination of the sediment may extend only to a depth of about 30 cm below the sediment surface. The upper 30 cm of contaminated sediments could then be treated differently from the remaining 70 cm of underlying "uncontaminated" sediments, reducing costs compared to special treatment of all the dredged sediment.

3.3 BOTTOM SEDIMENT SAMPLERS

The purpose of collecting the sample is to obtain an accurate representation of the nature of the sediment bottom in the study area. Therefore, the retained sample should resemble the original material as closely as possible without loss of a particular size or geochemical fraction. Disturbance or sample alteration can occur through sediment compaction, mixing, or fractional loss. Major sources of these disturbances are maneuvering of the sampling vessel in the shallow water prior to or during sampling; the pressure wave in advance of the lowered sampler; frictional resistance during sediment penetration by the sampler; tilting or skewed penetration of the sampler; and washout or other loss during retrieval to the sampling platform.

Regardless of the equipment chosen for sampling, it is useful to know the water depth at each station before starting the sampling. If this information is unavailable, it is recommended that the water depth be measured. Measuring equipment can range from a weighted chain to a highly accurate bathymeter. The purpose is to ensure adequate cable/rope length for operation of the correct equipment and to control the speed of entry of the sampler into the sediment. The speed of deployment of the sampler can be critical to good operation and sample recovery. Too rapid deployment generates and increases the shock wave advancing in front of the equipment. This shock wave can displace the soft unconsolidated surface sediments. Too fast deployment may also cause equipment malfunction, such as activating the trigger arm of a piston corer before achieving correct positioning.

It is also useful to have some understanding of the currents at the sampling site. Strong near-bottom currents can lead to poor equipment deployment, deflect a grab sampler, or require a long cable/wire to be deployed. Care should be taken to ensure that the weight of the sampler is adequate for working at the particular current conditions and that the sampler collects sediment at or very near the desired sampling site.

Table 3-1 Sediment Depth Collected by Different Samplers Under Optimal Conditions (about 2 m of fine-grained sediment)

Sediment depth sampled	Sampling equipment
0 to 10 cm	Lightweight, small volume grabs (for example, Birge-Ekman, Ponar and mini-Ponar, mini-Shipek)
0 to 30 cm	Heavy, large volume grabs (for example, Van Veen, Smith-McIntyre, Petersen)
0 to 50 cm	Single gravity corers (for example, Kajak-Brinkhurst and Phleger corers), box corers, multiple corers
0 to 2 m	Single gravity corers (for example, Benthos and Alpine)
Deeper than 2 m	Piston corers

Generally, there are two types of samplers (commercially available) used for collecting bottom sediments: grab samplers for collecting surface sediments, thereby providing material for the determination of the horizontal distribution of parameters; and corers for collecting a cross-section of sediments, thereby providing material for determination of the vertical distribution of parameters. The depth of the sediment collected by different surface sediment samplers and corers is given in Table 3-1.

The dimensions of individual grab samplers and corers described in this chapter are, in most cases, those of commercially available samplers. The actual dimensions of a particular sampler produced by different manufacturers may vary slightly. The following is a description of sediment samplers that the authors consider the most commonly used in studies of sediments.

3.3.1 Grab Samplers

Simplified drawings of commonly used grab samplers with their essential components are shown in Figure 3-9. Grab samplers consist either of a set of jaws that shut when lowered to the surface of the bottom sediment or a bucket that rotates into the sediment upon reaching the bottom. A large, vented top (usually a screen), or an opening in the back of the sampling bucket greatly reduces sediment disturbance caused by the shock wave in front of the descending sampler.

The properties of grab samplers that must be considered for general operational suitability are their stability and protection of the sample from washout. From the grab samplers described below, the Shipek and Ponar grabs have proved to be excellent general-purpose samplers capable of collecting most types of surface sediments. Both samplers maintain a near-perfect vertical descent and stable stance on the bottom in most waters with relatively weak currents, such as harbors and lakes. They are less suitable in fast-flowing rivers or the marine environment with strong currents. The use of small or lightweight samplers, such as the Birge-Ekman grab sampler, is advantageous because of easy handling, particularly from a small vessel, but becomes a disadvantage in areas with strong currents or during poor weather conditions with high waves

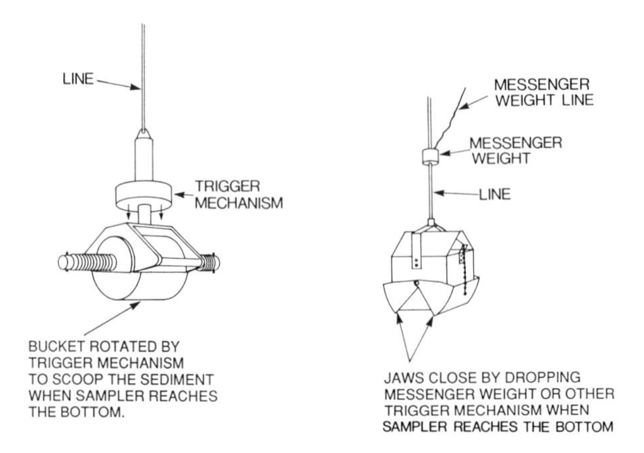

Figure 3-9 Grab samplers with their essential parts.

and intensive vessel motion. Under poor weather conditions, lightweight grabs are continuously lifted and dropped or are dragged along the bottom during sampling. The lightweight samplers are also less stable during sediment penetration and tend to fall to one side as a result of inadequate or incomplete penetration.

A surface layer of 2 to 3 cm of fine-grained, soft sediments can be lost due to washout. The loss of fine-grained sediments from a Shipek grab sampler during retrieval was investigated by Environment Canada's National Water Research Institute in Burlington, Ontario, by comparison of the concentrations of metals in surficial sediments collected by the Shipek grab sampler and a box corer in a depositional basin in Lake Ontario. The results indicated a loss of the topmost 2 to 3 cm of very fine, unconsolidated soft sediments from the Shipek grab sampler (Mudroch et al., 1988).

Samples collected by some grabs in firm, cohesive sediments, such as glacial till or glaciolacustrine clays, are often disturbed. For example, when the bucket of the Shipek grab sampler rotates into firm sediment, it cuts only a small sample filling about one-third to one-half of the bucket. The cohesive sediment has sufficient space in the bucket to turn as one piece upside down during the ascent of the sampler from the bottom, making recognition of the sample's surface difficult upon retrieval.

Sample volume also needs to be considered when choosing a proper surface sediment sampler. Grabs that can collect more material, such as the Petersen grab sampler, are favored for sampling in biological studies requiring a large sample volume.

Following is a description of the basic, most commonly used and available surface sediment samplers. The choice usually depends on the sampled sediment depth and volume, handling suitability under given sampling conditions, and, very often, personal preference. For example, three surface sediment samplers, the Shipek and Ponar grabs and Birge-Ekman sampler, are most commonly used by personnel at Environment Canada's National Water Research Institute, Burlington, Ontario, for collecting surface sediments from Canadian inland waters. The larger, heavier Van Veen or Ponar grab samplers are more commonly used for collection of surface sediments in the Canadian marine environment. Some of the commercial sediment samples were modified in a specific use to help achieve the objectives of the sediment sampling program. More details and descriptions of different grab samplers are given by Mudroch and MacKnight (1994).

Birge-Ekman Sampler — Petite (Standard Size in Parentheses)

Sampled area: 15 cm × 15 cm (23 cm × 23 cm)
Cutting height: 15 cm (23 cm)
Sample volume: 3,400 cm^3 (13,300 cm^3)
Weight: empty about 5 to 10 kg, depending on the material (13 kg)
Weight with the sediment: 10 to 15 kg (40 kg)

The Birge-Ekman sampler (Figure 3-10) is available in several sizes with sample chambers ranging from 3,500 to 28,320 cm^3 and with carrying cases. A tall version is also available with optional weights for deeper penetration into sediment and a 1.5-m long operating handle for shallow-water applications. The standard Birge-Ekman sampler can be operated manually. For the larger models a winch or crane hoist is recommended. All models are composed of a stainless steel or brass box with a pair of jaws and free-moving, hinged flaps. The spring-tension, scoop-like jaws are mounted on pivot points on opposite

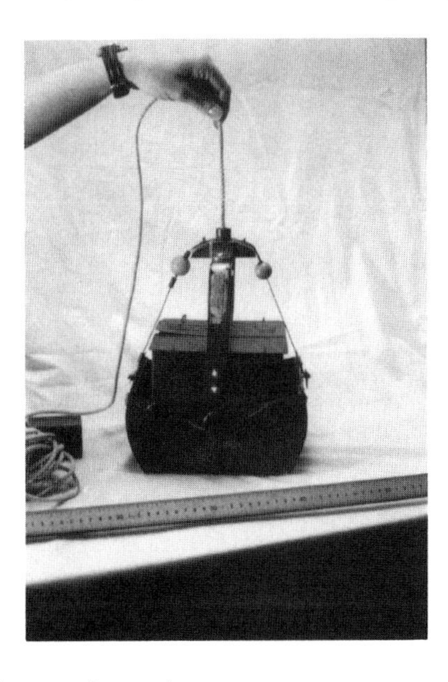

Figure 3-10 Birge-Ekman grab sampler.

sides of the box. The jaws are held open by stainless steel wires that lead to an externally mounted trigger assembly, activated by a messenger. After closure, the jaws meet tightly along the seams to prevent washout during retrieval. A weighted messenger is secured to the sampler's line so it can freely move along the line and not become lost during sampling. It is highly advisable to have a spare messenger weight on board the sampling vessel. When the sampler has reached the sediment the messenger is sent down the rope or wire to activate the jaw closure mechanism. During descent through the water, the flaps are forced open by the pressure of water passing through the open-jawed box. The flaps cover the surface of the box during retrieval of the sample, preventing disturbance of the collected sediment.

The Birge-Ekman sampler is suitable for sampling fine-grained, soft sediments and a mixture of silt and sand. Larger objects (i.e., gravel, shells, macrophytes, or pieces of wood) trapped between the jaws will prevent jaw closure and result in sample loss. Due to its weight and the need to use an activating messenger, the sampler must be used under low-current conditions and penetrate perpendicular to the sediment. In very soft sediments with a high water content, the sampler tends to penetrate too deeply due to its weight. This can be prevented by dropping the messenger weight immediately after the sampler reaches the bottom. With careful deployment, a small but relatively undisturbed sample can be obtained. Upon retrieval, the sediment can be divided into several subsamples through the flaps at the top of the sampler. However, to empty the sampler quickly, sediment has to be removed through the bottom by opening the jaws over a container. In the latter case, the sample has to be treated as a bulk surface sediment sample. Reliability of soft sediment sampling has been improved by the modification of the standard Ekman grab sampler (Håkanson, 1986).

Ponar Grab Sampler — Standard (Petite Size in Parentheses)

Sampled area: 23 cm × 23 cm (15 × 15 cm)
Weight: about 23 kg (10 kg)
Maximum sample volume: 7,250 cm^3 (1,000 cm^3)
Required lifting capacity: 100 kg

The Ponar grab sampler (Figure 3-11) is used with a winch or crane hoist. It consists of a pair of weighted, tapered jaws held open by a catch bar across the top of the sampler. On touching the bottom, the tension on the bar is released, allowing the jaws to move and close. A special mechanism of the Ponar grab sampler prevents accidental closing during handling or transport. The device is activated by the release of the cable/rope tension on the lifting mechanism when the sampler reaches the sediment. During retrieval, the tension on the cable keeps the jaws closed. The sampler has to be lowered slowly under the water surface to avoid premature triggering on impact with the water surface. When in water, the sampler can be lowered until approximately 5 m from the bottom, then it must be lowered slowly. A steady, slow

Figure 3-11 Ponar grab sampler (regular and petite size).

winch speed should be maintained to lift the sampler from the bottom after its penetration of the sediment. The jaws of the sampler overlap to minimize sample washout during ascent of the equipment. The upper portion of the jaws is covered with a mesh screen and rubber flap, allowing water to pass through the sampler during descent, reducing disturbance at the sediment-water interface by a shock wave. Upon recovery, the mesh screen can be removed, providing easy access to the recovered sediment for subsampling. Where a bulk sediment sample is required, or the entire sediment sample needs to be sieved for benthic organisms, it is easy to empty the sampler by opening its jaws over a sufficiently large container. The Ponar grab sampler has a pair of metal side plates that prevent the sampler from falling over after jaws closure, reducing washout and helping to preserve a very good sediment sample. This sampler is suitable for sampling most sediment types from soft, fine-grained to firm, sandy material, with the exception of hard clay, in both freshwater and marine environments with little or no disturbance.

Commercially available, the petite Ponar grab sampler (weight about 10 kg, sampled area 15 cm × 15 cm, sample volume 1,000 cm^3) is designed for hand line operation, but its construction and operation are similar to the standard Ponar grab sampler. Similar to the standard Ponar grab sampler, the petite Ponar grab sampler is very good for sampling coarse and firm bottom sediments.

Van Veen Grab Sampler — Standard (Large Size in Parentheses)

Sampled area: 35 cm × 70 cm (50 cm × 100 cm)
Weight: 30 kg (65 kg); with weights, 40 kg (85 kg)
Sample volume: 18 l (up to 75 l)
Required lifting capacity: 150 to 400 kg

The Van Veen grab (Figure 3-12) is suitable for obtaining bulk samples ranging from soft, fine-grained to sandy material for biological, hydrological, and environmental studies in deep water and strong currents in the marine environment. The Van Veen grab sampler is manufactured in several sizes from hot-galvanized steel or stainless steel, which is particularly suitable in pollution studies. The weighted jaws, chain suspension, and doors and screens allow flow-through during lowering to the bottom and assure vertical descent where strong underwater currents exist. The relatively large surface area and the strong closing mechanism allow the jaws to excavate relatively undisturbed sediments. A shock-wave is not created when the sampler settles on the bottom. As the lowering wire slackens, the hook on the release device rotates and the short suspension chains fall free. When the wire is slowly made taut, the chains attached at the top of the release exert great tension on the long arms extending beyond the jaws, causing them to lift, dip deeper into the sediment, and trap material as they tightly close. The stainless-steel door screens have flexible rubber flaps which, during lowering, are lifted. When the grab settles on the bottom, the flaps fall back and cover the screens completely, preventing any loss of sediment during retrieval. Catches on the jaws are provided to lock the doors. When the grab is on the sampling platform, the sediment sample may be dumped into a box or container for further handling. Alternatively, the doors can be unlocked and the samples collected before dumping. The Van Veen grab sampler is similar in operation to the Petersen or Ponar grab samplers, but tends to be larger and heavier, has top access doors for sub-sampling, and has no internal parts to contaminate the sample, such as, the chain in the Peterson grab sampler. Because of its typical weight and size, the Van Veen grab sampler is more commonly used in the marine environment, where there is deeper water and strong currents.

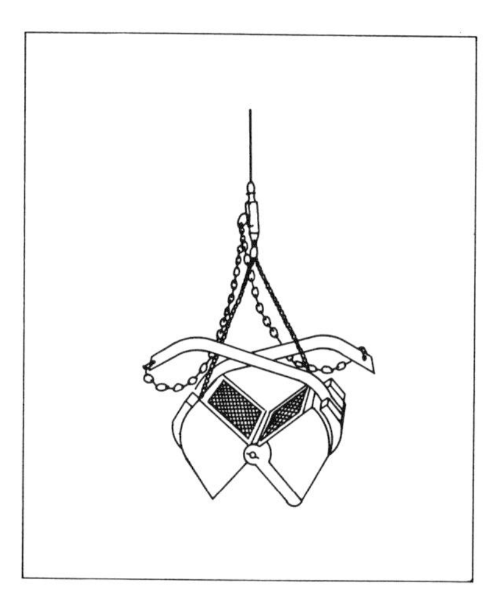

Figure 3-12 Van Veen grab sampler.

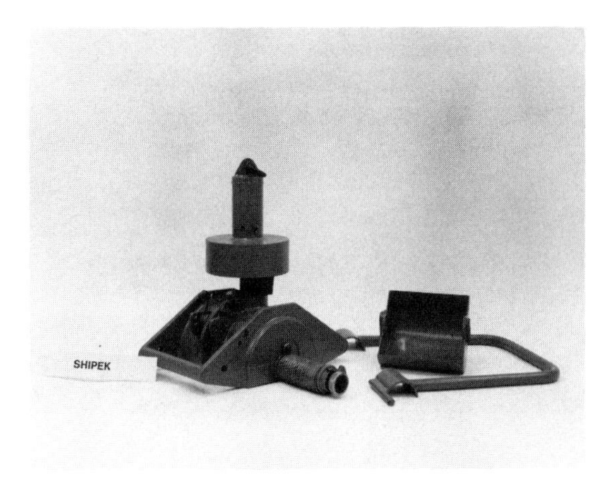

Figure 3-13 Shipek grab sampler with the sampling bucket and cocking lever.

Shipek Grab Sampler

Sampled area: 20 cm × 20 cm
Weight: 50 kg
Sample volume: 3,000 cm^3
Required lifting capacity: 200 to 300 kg

The Shipek grab sampler (Figures 3-13) was designed to obtain relatively undisturbed samples of sediments ranging from soft, fine-grained to sandy material, even from sloping bottoms, from any depth. The sampler operates by spring-driven rotation of a bucket upon contact with the bottom, encircling the sediment to about 10 cm depth. It consists basically of a steel shank, weight, and bucket. Of the two concentric half-cylinders, the inner half-cylinder is rotated at high torque by two helically wound external springs. The sampler is lowered with the bucket in an inverted position until it contacts the sediment surface. An internal weight triggers a release mechanism. The bucket is then forced to rotate on its axes at high speed by two helical springs. It cuts through the sediment and at the conclusion of a rotation of 180° is stopped and held in an upwards, closed-concave position. Cast into each end of the sampler frame are large stabilizing handles.

The Shipek grab sampler should be lowered slowly until the trigger weight is submerged. The lowering velocity should not exceed the terminal velocity of approximately 100 m/min. The sampler must be raised cautiously from the bottom. Retrieval rates up to 200 m/min can be used. On the sampling platform the bucket is released by pulling outwards on the pivot pins and supporting the bottom of the bucket with both hands. Washout can occur during the ascent of

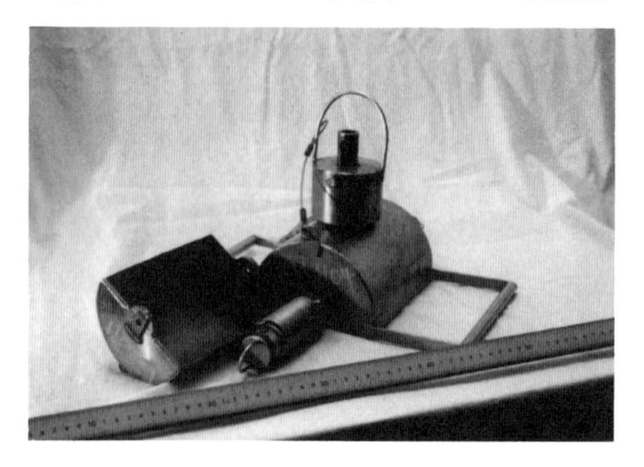

Figure 3-14 Mini-Shipek grab sampler.

the sampler from the bottom, particularly if the uppermost material is soft and fine-grained. Sampling of hard sediments, such as firmly packed sand and glaciolacustrine clay or till, can be unsuccessful and the samples can be tilted or disturbed. There is easy access to the sample when the bucket is removed from the holding system. This enables visual observation and description of the individual sediment layers, and horizontal or vertical subsampling from the bucket. Larger objects, such as gravel, pieces of wood, shells, etc., trapped between the edge of the bucket and the body of the sampler can cause disturbance and washout of the sample. They should be removed only by rotating the bucket into an open position, but they should never be pulled out of the sampler by hand. The triggering mechanism is very sensitive, and extreme caution is necessary when the bucket is rotated inside the sampler and is ready for lowering to the bottom.

A double-Shipek grab sampler consists of two single Shipek grab samplers joined by metal bars. It is suitable for obtaining duplicate samples at the sampling site. However, the weight and sufficient space at the sampling platform for handling the double-Shipek have to be considered.

A mini-Shipek grab sampler, (weight about 5 kg, sampled area 10 cm × 15 cm, sample volume up to 500 cm^3), constructed at Environment Canada's Canada Centre for Inland Water, Burlington, Ontario (Figure 3-14), operates on the same principle as the standard Shipek grab sampler. The sampler is most suitable for collecting fine-grained, soft sediments. In sand and other firm material, such as glaciolacustrine clay or till, only up to 3 cm of surficial sediment is recovered due to the light weight and small size of the sampler. In addition, the sample from a hard bottom may be tilted and will not represent an undisturbed surface sediment. To obtain a good-quality sample, the mini-Shipek grab sampler has to penetrate the sediment surface vertically. A consistent, slow lowering speed of the sampler is necessary to prevent the sampler from triggering before reaching the bottom. Washout of very fine material from the surface of the sample is very likely to occur during the retrieval, similar to the standard Shipek grab sampler. Because of its weight, the mini-Shipek grab sampler is suitable for hand-line operation from different sampling platforms.

3.3.2 Corers

Corers are fundamental tools for obtaining sediment samples for geological and geotechnical survey and, recently, for the investigation of historical inputs of contaminants to aquatic systems. Generally, corers consist of a hollow metal pipe (or a plastic pipe) that is the core barrel, varying in length and diameter; easily removed plastic liners or core tubes that fit into the core barrel and retain the sediment sample; a valve or piston mounted on the top of the core barrel that is open and allows water to flow through the barrel during descent, but shuts upon penetration of the corer into the sediment, thereby preventing the sediment from sliding from the corer during the ascent; a core catcher to retain the sediment sample; a core cutter for better penetration of the sample; removable metal weights (usually lead coated with plastic) to increase penetration of the corer into the sediment; and stabilizing fins to assure vertical descent of the corer. Typical parts of a corer are shown in Figure 3-15.

The cutter is mounted on the end of the core barrel to achieve better and deeper penetration into the sediment. Commercially available cutters are typically made of stainless steel or steel, and have screw, bayonet, or setscrew mounts. Brass or plastic cutters have also been used. The cutting edge should be easy to sharpen when it becomes dull or damaged by gravel or other materials in the sediment. A core catcher is inserted inside the cutting head of the corer to prevent loss of the sediment during retrieval. The sample is retained in the liner by a series of spring-loaded metal fingers that allow the sediment to enter, but not to fall out of, the liner. A core catcher is less effective in soft sediments with a high water content than in consolidated ones. It also "rakes" the sample during entry, disturbing sections of the core. There is also a

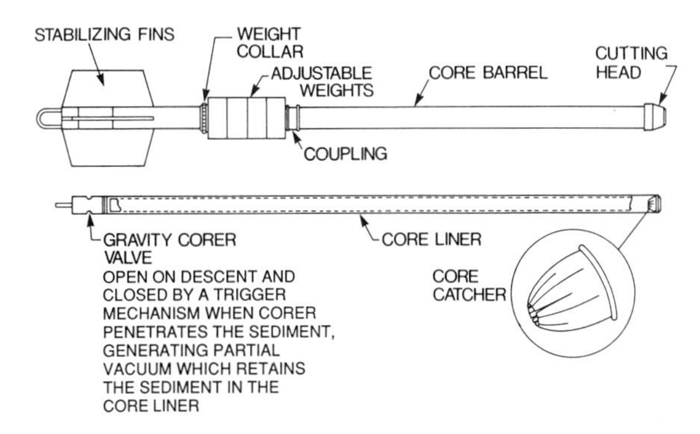

Figure 3-15 Typical parts of a sediment corer.

Table 3-2 Estimated Weight of Dried Sediment Layer Subsampled from Core Liners of Different Diameters

	Approximate weight (g) of dry material in uppermost 1-cm sediment layer[a]	
Tube inner diameter (cm)	Soft fine-grained sediment[b]	Firmer silty clay[c]
3.5	0.7 to 1.4	1.1 to 2.2
5.08	1.5 to 3.0	2.3 to 4.6
6.6	2.6 to 5.2	3.2 to 7.9
10.0	5.9 to 11.8	8.8 to 17.7

[a] Based on 90 to 95% sediment water content.

[b] Based on specific gravity of sediment 1.5 (high organic matter content).

[c] Based on specific gravity 2.3 (low organic matter content).

possibility of contamination from lubrication of the spring that is a part of some core catchers. Plastic core catchers can be used but have a poor ability to rebound to their former shape. Different types of core catchers were reviewed by Bouma (1969).

Different dimensions of core tubes (metal tubes or plastic liners) used with the corers affect the quantity of the recovered sediment sample. The sample size is important, particularly when more parameters need to be determined in the sediment sample. An important part of the core, the uppermost 1- to 3-cm layer, often consists of soft material with a high water content. Estimated weights of an uppermost 1-cm layer of sediment subsampled from core liners of different inner diameters are given in Table 3-2.

The quantity of the sediment in a core tube below the topmost 1-cm layer gradually increases due to the lower water content and sediment compaction. In fine-grained material, the water content is about 80% at the 10-cm sediment depth, 70% at the 20-cm sediment depth, and about 50 to 60% at the 30- to 40-cm sediment depth. Below 50 cm, the sediment usually becomes more compacted and there is little change in the water content. An exception can be sediment from small lakes with high organic matter content and restricted water circulation. Water content in this type of sediment can be 90 to 95% at the 50- to 100-cm sediment depth, and sampling will require a special coring device to retain the sediment in the core tube during retrieval.

Several types of coring devices have been developed for sediment sampling:

- Single gravity corers featuring a core barrel penetrating the sediment by gravity and collecting up to 2 m of sediment.
- Multiple gravity corers, featuring two to four core barrels.
- Box corers for collecting a rectangular sample from the upper-50-cm sediment layer.
- Piston corers featuring a core barrel with a liner and piston for collecting cores 20 m and longer in deep water.
- Boomerang corers for taking samples from the sea floor.

- Vibracorers featuring a vibrating device and a stationary piston for collecting samples from hard clays, shales, and recent calcareous sandstones (Rosfelder and Marshall, 1967).

Commercially available coring equipment ranges from a small, hand-operated corer that can be used in shallow water, to a heavy, oceanographic core sampler.

The shock wave produced during the free-fall of most corers can disturb the surface sediment. Open core barrels and core valves with unrestricted water flow allow water to pass freely through during lowering of the equipment to the bottom and limit the shock wave. The degree of disturbance appears to be a function of the speed of impact and the surface area of the core. With the decrease in speed of entry of the corer into the sediment, the sediment disturbance becomes smaller, but there is also less penetration and, consequently, a shorter core is recovered.

During sediment penetration by a corer, frictional resistance results in sediment deformation and compaction. Using a corer with thin, smooth walls and sharpening the lower end of the core tube to a small angle reduces frictional resistance.

A corer may enter at an angle not perpendicular to the bottom or, in the worst case, may plunge sideways into the sediment. This usually occurs when the vessel drifts during sampling or the corer is lowered too quickly. If the upper sections of the coring device emerge covered with sediment upon sample retrieval, there is a strong likelihood that the corer tilted in the sediment because the speed of entry was too fast or it encountered a firm underlying material, such as compacted sand or rock. A sampler covered with sediment also may be a sign of over-penetration by high-speed entry into fine-grained sediments. The sloping surface of the sediment recovered in a core tube indicates that the corer penetrated at an angle.

The degree of disturbance that can be tolerated in a sediment core varies with the intended use of the results obtained. Studies of sediment contaminants are primarily concerned with profiling the concentrations of recently deposited contaminants in the sediments. Assuming an annual rate of sediment deposition of 0.1 cm/year, the concentration of contaminants in the uppermost 1 cm sediment layer indicates the input of contaminants for the investigated area during the past 10 years. The loss of a few millimeters of the surface sediment will cause underestimation of the contaminant loadings to the site.

Sediment cores with a minimum of disturbance are required for continuous long-term monitoring of contaminant concentrations in sediments carried out to assess the efficacy of remedial actions in the drainage basins of rivers and lakes, and to determine the rate of natural burial of contaminants in sediments by deposition of clean material. For example, a new waste treatment technology was implemented at a plant discharging its effluent into a stream entering a lake in a remote area. The efficacy of the new technology and the burial of contaminated sediments may be assessed by sampling and analyzing sediments collected at depositional areas in the lake over the next 15 to 20 years. The changes in the concentrations of contaminants in the sediment column will reflect the changes of contaminant concentrations in the effluent and indicate the rate of burial of contaminated sediments. Consequently, sampling an undisturbed sediment column will be required during the 15 to 20 years of monitoring.

Hand Corers

Most hand corers are suitable for collecting soft or semi-compacted sediment samples by hand in marshes, tidal flats, rivers, and other shallow water areas, or in deep water by a diver. Commercially manufactured models usually consist of a metal or plastic core tube 3.5 to 7.5 cm I.D. and extension handles on the top end for driving the corer into the sediment. Colored plastic caps should be used for sealing and identifying the top and the bottom of the core tube or the liner. Commercially supplied hand corers can have extra handles of various length (about 1 to 5 m) which, when attached to the hand corer, allow the collection of samples from depths equal to the length of the handle. The core tubes and liners are usually 50 to 120 cm long, threaded on both ends, and tapered on the bottom for easier penetration of the sediment. A nose-piece can be attached at the bottom and, if found advantageous, a core catcher can also be installed. The weight of hand corers varies from 5 to 17 kg; the extension handles add another 4 to 12 kg.

Single-Gravity Corers

There are few corers which can be operated without a mechanical winch. Kajak-Brinkhurst and Phleger corers and their modifications are the representatives of this group. Surface sediment sampling from a small vessel can be a single-person operation; however, at least two operators are required to stabilize the vessel at the sampling station, to lower the corer overboard, and cap the bottom of the retrieved core. The Benthos Gravity Corer and the Alpine Gravity Corer, also described in this section, are winch- or crane-deployed corers. In addition to these corers, there are several others available from commercial suppliers of aquatic sampling equipment.

Kajak-Brinkhurst (K-B) Corer

Weight: about 9 kg (standard size), lead weight 7 kg
Core tube size: 5 cm I.D.
Core tube length: 50 cm, 75 cm

The corer (Figure 3-16) is suitable for sampling soft, fine-grained sediments and recovering up to about 70-cm long cores. The standard K-B corer is a messenger-operated sampler with unrestricted water flow during its descent. The corer is suitable for sampling soft, fine-grained sediments. The

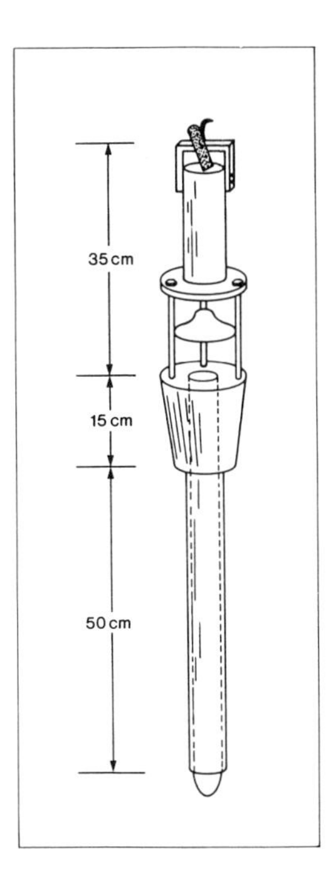

Figure 3-16 Kajak-Brinkhurst corer.

closure of the messenger-operated valve allows the operator to choose the closing time of the valve when he feels that the sampler has sufficiently penetrated the bottom sediment. Closing the valve by the messenger creates a partial vacuum inside the core tube during the ascent of the sampler from the bottom, and assists in the retention of the sediment in the tube. The K-B corer was improved by the addition of an automatic trigger mechanism to replace the messenger. The major problem in using the messenger was that after the corer vertically penetrated the sediment, the line was often "streaming," i.e., was not vertical from the water surface to the corer, in which case the messenger would not properly activate the valve. There are various types of this corer commercially available with various accessories, such as stabilizing fins, extra weights, and various core tubes (PVC, Lexan, brass). The standard K-B corer can be operated manually, but a winch is recommended for a "heavy" K-B corer with accessories and 75 -cm long core tubes. A 5-cm I.D. core tube used with this corer recovers a greater quantity of sediment than the 3.5-cm I.D. core tube used with the Phleger corer described below.

Phleger Corer

Weight: about 8 kg (without lead weights; additional 7 kg per each added lead weight)
Core tube size: 3.5 cm I.D.

The Phleger corer (Figure 3-17) is suitable for sampling different types of sediment ranging from soft to sandy, semi-compacted material, as well as peat and vegetation roots in shallow lakes or marshes. The length of the obtained core is up to 50 cm. A relatively narrow, 3.5 cm I.D. core liner recovers a small quantity of material, which is a disadvantage, particularly when the core needs to be subsampled into small sections. The core barrel has a bayonet fitting nose cutter. The upper part of the core barrel screws into a further section of tubing on which ring weights are mounted. The upper tubing supports the weight rings and provides excellent vertical stability during core descent. The upper tube is capped with a valve assembly consisting of a neoprene bung mounted on a

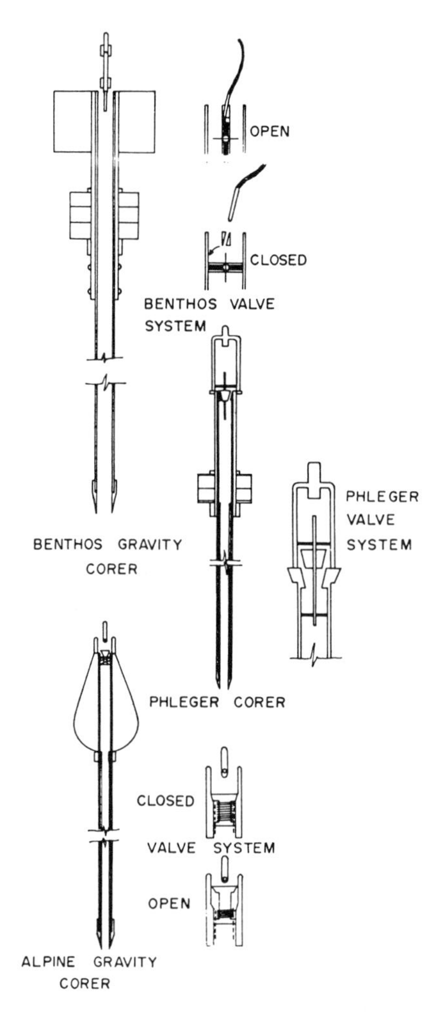

Figure 3-17 Phleger, Benthos, and Alpine gravity corers with their valves.

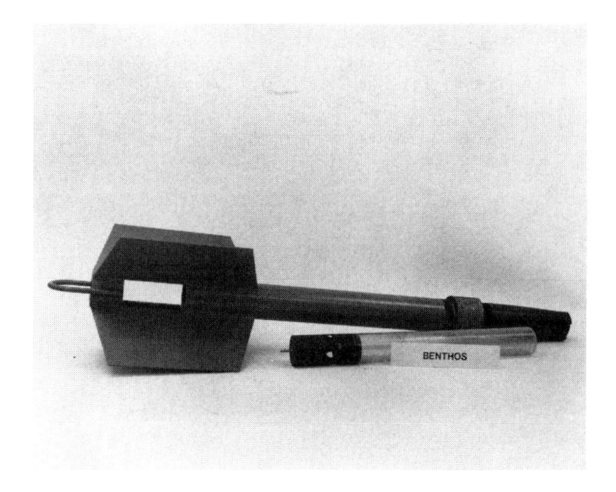

Figure 3-18 Benthos gravity corer.

metal pin that slides in two locations. The bung is slightly tapered and fits into a similarly shaped metal seating. The water pressure within the tube forces the bung up and clear of its seat on lowering and penetration. On withdrawal, the pressure is maintained within the tube by the bung as it slides back into its seat and seals perfectly, thereby retaining the sediment sample.

Benthos Gravity Corer

Weight: 25 kg and up to 6 × 20 kg of additional lead weights
Core tube size: 6.6 cm I.D., 7.1 cm O.D.
Required lifting capacity: 350 to 500 kg

The Benthos gravity corer (Figures 3-17 and 3-18) was designed to recover up to 3 m long cores from soft, fine-grained sediments. On the recent model stabilizing fins on the upper part of the corer promote vertical penetration into the sediment. To enhance penetration, up to six 20-kg weights can be mounted externally on the upper part of the metal barrel. A valve system at the top of the liner prevents loss of the sample from the tube. The valve is fitted to the top of the core liner, which is then inserted into the core barrel. The valve is a critical part of the corer and the success of sampling depends on its proper operation. This led to various designs of the valve by the corer manufacturing company. For example, the valve presently used at Environment Canada's National Water Research Institute, Burlington, Ontario, is an auto-valve held open by the water flow during descent and penetration. Upon retrieval, the suction of the sediment attempting to slide out of the core tube and the force of the spring push the plunger into a machined seat. The created vacuum holds the sediment in the tube. The valve should always be carefully sealed in the liner and its operation regularly inspected.

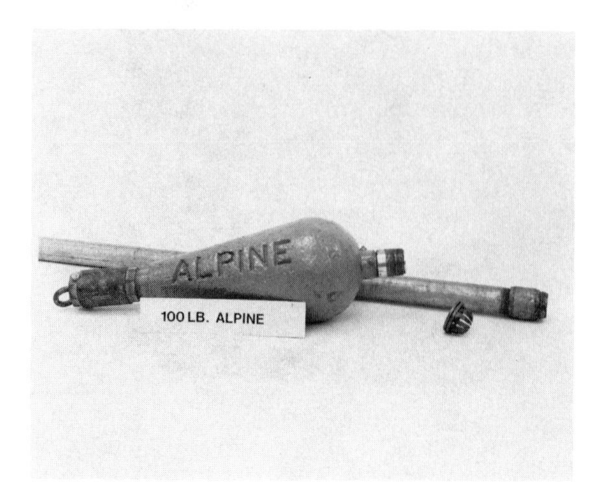

Figure 3-19 Alpine gravity corer.

Alpine Gravity Corer (model 211)

Weight: 110 kg
Plastic liners: 3.5 cm I.D., 3.8 cm O.D.
Required lifting capacity: 500 kg

The Alpine gravity corer (Figures 3-17 and 3-19) is finless and has an interchangeable steel barrel (4.1 cm I.D., 4.8 cm O.D.) in lengths of 0.6, 1.2, and 1.8 m. A weight on the corer is included in the barrel. Attached to the top of this is a combination attachment point/valve assembly. The valve system uses a light compression spring to retain a plastic and rubber leg and cap assembly against a bevelled, circular seat. During penetration, the increased pressure in the barrel causes the cap assembly to lift off its seat and allow the necessary displacement of water from the barrel. Penetration and pressurized displacement cease simultaneously. The compression ring then forces the cap valve to retreat and seal prior to withdrawal.

Sly (1969) tested an Alpine gravity corer with extremely variable results. The most successful cores were obtained by allowing free-fall for a distance up to twice the barrel length used. However, due to the lack of fins, vertical penetration of the corer was not obtained in many cores. The worst entry observed was approximately 25° from the vertical. Under good working conditions the angle reduced to 5°, and the corer embedded itself deeply in the mud. Sheared laminae and disturbed surfaces were observed on radiographs of hundred of cores.

Multiple-Gravity Corers

Multiple corers typically consist of several core barrels mounted on a single fin and weight system. They have been developed for multiple sampling at one site, comparative studies, evaluation of sediment sampling precision, and determination of sediment heterogeneity over a small area.

For example, a triple Benthos corer, built at the Canada Centre for Inland Waters, is based on the same operational principles as the single Benthos gravity corer. It has an outer ring that houses three evenly spaced core barrels welded to a fourth nonfunctional barrel. The valves and trigger mechanism are original Benthos products. The core tubes are 50 cm in length and have an outside diameter of 7.1 cm.

Brinkhurst et al. (1969) described a multiple unit of the K-B corer and its application in studies of sediment biota. Other different multiple corers were described by various scientists (Hamilton et al., 1970; Kemp et al., 1971; Jones and Watson-Russell, 1984).

Box Corers

Box corers are gravity corers that were developed in the late 1950s, and later modified and refined to improve their operation (Bouma and Marshall, 1964; Bouma, 1969; Bruland, 1974; Sundby et al., 1981; Mawhinney and Bisutti, 1987). They were designed for collecting large rectangular sediment cores in biological and geological studies at various water depths, variable penetration rates, and different sediment types. There are two basic designs to the bottom mechanism of the box corer: (1) the Ekman design, in which there are two bottom flaps that can be triggered and closed much like the Ekman grab sampler; and (2) the Reineck design, in which a large shovel-like device is activated and slides across the bottom of the box corer. There are several box corers of different design and size commercially available. Recognition of the excellent quality of the undisturbed sediment samples collected by box corers, particularly in studies of the sediment-water interface, initiated design and construction of different custom-made box corers for special requirements.

Generally, a box corer consists of a stainless steel box of a variable size. Most box corers are equipped with a frame that also ensures vertical penetration on low slopes and stabilizes the sampler on the bottom. Due to its heavy weight (up to 800 kg) size (up to 2 m × 2 m), and required lifting capacity (on the order of 2,000 to 3,000 kg), the box corer can be operated only from a vessel with a large lifting capacity and sufficient deck space. Unless specially designed, the core retainer of a box corer can be damaged by penetration into a thin layer (less than 30 cm) of soft, fine-grained sediment underlain by firm material containing gravel or boulders. Therefore, such box corers should be used only in areas with a minimum 1-m layer of soft, fine-grained sediments. Box corers are triggered automatically when they reach the bottom. However, the actual sediment coring is carried out after the device is on the bottom.

The box corer described below and shown in Figures 3-20 to 3-22, was redesigned and custom-made using the design of the box corer described by Bouma (1969) and is used at Environment Canada's National Water Research Institute, Burlington, Ontario. The box corer consists of a number of basic

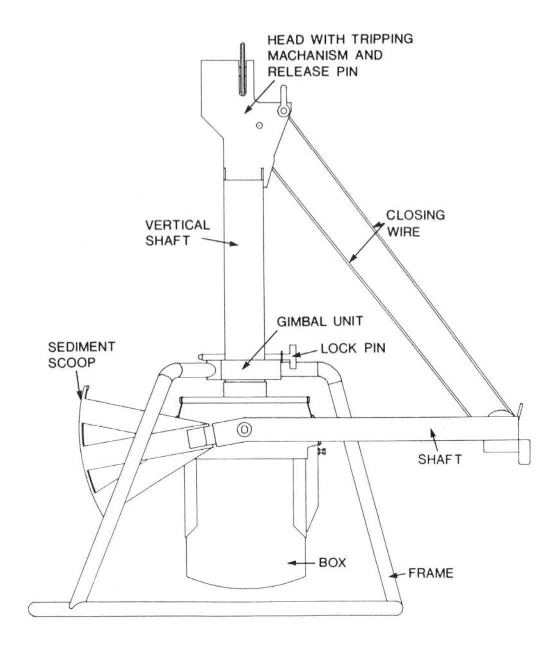

Figure 3-20 Box corer.

Figure 3-21 Deployment of the box corer.

parts: gimballed frame, control stem with two box holders, closing mechanism, tripping mechanism, and sampling box. The central frame slides through the gimballed top of the frame in such way that samples are always taken vertically. The closing mechanism consists of a blade at the end of a double arm, which pivots about the box holder. A tripping mechanism on top of the central stem makes it possible to use only one wire for lowering, sampling, closing, and returning to the surface. Very little free-fall is possible, allowing penetra-

Figure 3-22 Lowering the box corer with retrieved sediment to the ship deck.

tion into the sediment primarily based on gravity. Weights can also be added to the central stem by removal of a plate on the side to allow deeper penetration of the sediment. The hollow pipe frame ensures vertical penetration at slopes up to 18° and prevents the sampler from falling over on the bottom.

The sediment inside the box corer can be subsampled by inserting core tubes into the sediment (Figure 3-23). The top of the core tubes have to be sealed by core caps to prevent the sediment from falling out during recovery of the cores from the box corer. Since all subsamples come from the same small area of the bottom, the results of any analyses and studies of the sediment carried out at various laboratories can be compared. The lack of disturbance of the sediment recovered in the box corer can be ascertained by the sediment appearance. Hand-coring of the sediment from the box corer allows good control of the compaction of the sediment upon driving the core tubes into the sediment, particularly when clear plastic tubes are used.

Carlton and Wetzel (1985) described a box corer that permitted collection of an undisturbed section of unconsolidated sediment (cross-sectional area >700 cm^2) with overlying water. The design of the sampler allowed the collected sample to be held as a laboratory microcosm under the original *in situ* or other controlled condition. It was suggested that the sampling technique should provide samples appropriate for numerous applications, such as studies of biogeochemical cycling and solute flux or different bioassays with benthic populations. Recently, a box corer was developed that enables horizontal subsampling of the entire sediment volume recovered by the box corer (Brunskil, 1989).

Piston Corers

Piston corers are usually used in studies of bottom sediment stratigraphy in oceans and deep, large lakes and can typically recover relatively undisturbed

A

B

Figure 3-23 Subsampling of sediment in a box corer by hand-coring (A) and capping the inserted core tubes (B).

cores of 3 m in length, or up to about 20 m long. The large weight and size of the piston corer require a large vessel with heavy-duty cranes with a lifting capacity over 2,000 kg, and experienced operators. Figures 3-24 and 3-25 show the operation of a piston corer sampling Lake Ontario sediments. The piston corer is not suitable for sampling a sediment profile in studies of contaminants because of the disturbance of the topmost layer of the sediment by the corer. Sediment sampling by piston corers in different studies was summarized by Mudroch and MacKnight (1994).

Vibratory Corers

Typically, corers can only penetrate a few centimeters into sandy unconsolidated sediments. Vibratory corers overcome the resistance factor by a

Figure 3-24 Preparation of the piston corer for sediment coring.

Figure 3-25 Cleaning the piston corer pipes before lifting the corer to the ship deck.

vibration action of the core barrel (either side to side or up and down). Vibratory coring systems are used mainly to assess geotechnical or structural properties of the sediment. They have rarely been used for obtaining sediment samples for the study of environmental pollution. Different vibrating corers and their use were described by Sly and Gardener (1970), McMaster and McClennen (1973), Dokken et al. (1979), Fuller and Meisburger (1982), and others.

REFERENCES

Bascom, W.N., Instruments for measuring pollution in the sea, *Prog. Water Technol.*, 4, 99, 1979.

Bouma, A.H., *Methods for the Study of Sedimentary Structures*, John Wiley & Sons, New York, 1969, 458.

Bouma, A.H. and Marshall, N.F., A method for obtaining and analyzing undisturbed oceanic sediment samples, *Mar. Geol.*, 2, 81, 1964.

Brinkhurst, R.O., Chua, K.E., and Batoosingh, E., Modification in sampling procedures as applied to studies on the bacteria and tubificid oligochaetes inhabiting aquatic sediments, *J. Fish Res. Board Can.*, 26, 2581, 1969.

Bruland, K.H., *^{210}Pb Geochronology in the Coastal Marine Environment*, Ph.D. thesis, University of California, San Diego, 1974.

Brunskil, G.J., personal communication, 1989.

Carlton, R.G. and Wetzel, R.G., A box corer for studying metabolism of epipelic microorganisms in sediment under *in situ* conditions, *Limnol. Oceanogr.*, 30, 422, 1985.

Collins, V.G., Methods in sediment microbiology, in *Advances in Aquatic Microbiology, Vol. 1*, Droop, M.R. and Jannasch, H.W., Eds., Academic Press, New York, 1977, 219.

Dokken, Q.R., Circe, R.C., and Holmes, C.W., A portable, self supporting, hydraulic vibracorer for coring submerged, unconsolidated sediments, *J. Sed. Petrol.*, 49, 658, 1979.

Elliot, J.M. and Tullett, P.A., *A Bibliography of Samplers for Benthic Invertebrates*, Freshwater Biological Association, Occasional Publication No. 4, 1978.

Elliot, J.M. and Tullett, P.A., *A Supplement to a Bibliography of Samplers for Benthic Invertebrates*, Freshwater Biological Association, Occasional Publication No. 20, 1983.

Fowler, G.A. and Kingston, P.F., The use of bottom operating sampling equipment for investigation of the continental shelf, in *Proc. 1st Can. Conf. Mar. Geotech. Engineering*, 1979, 238.

Fuller, J.A. and Meisburger, E.P., A simple, ship-based vibratory corer, *J. Sed. Petrol.*, 52, 642, 1982.

Håkanson, L., Modifications of the Ekman sampler, *Int. Revue. Ges. Hydrobiol.*, 71, 719, 1986.

Håkanson, L. and Jansson, M., *Principles of Lake Sedimentology*, Springer-Verlag, Berlin, 1983.

Hamilton, A.L., Burton, W., and Flannagan, J.F., A multiple corer for sampling profound benthos, *J. Fish Res. Board Can.*, 27, 1867, 1970.

Holme, N.A., Methods of sampling the benthos, in *Advances in Marine Biology*, Vol. 2, Russel, F.S., Ed., Academic Press, New York, 1964, 171.

Hopkins, T.L., A survey of marine bottom samplers, in *Progress in Oceanography*, Vol. 2, Sears, M., Ed., Pergamon-MacMillan, 1964, 213.

Jones, A.R. and Watson-Russell, C., A multiple coring system for use with scuba, *Hydrobiologia*, 109, 211, 1984.

Journal of Glaciology, Instruments and methods, ice drills and corers, *J. Glaciol.*, 3, 30, 1958.

Kajak, Z., Benthos of standing water: survey of samplers, in *A Manual on Methods for the Assessment of Secondary Productivity in Fresh Water*, IBP Handbook No. 17, Edmonson, W.T., Ed., Blackwell Scientific, Oxford, 1969, 25.

Kemp, A.L.W., Saville, H.A., Gray, C.B., and Mudrochova, A., A simple corer and method for sampling the mud-water interface, *Limnol. Oceanogr.*, 16, 689, 1971.

King, E.W. and Everitt, D.A., A remote sampling device for under-ice water, bottom biota, and sediments, *Limnol. Oceanogr.*, 25, 935, 1980.

Le Tirant, P., Seabed exploration coring devices and techniques, in *Seabed Reconnaissance and Offshore Soil Mechanics for the Installation of Petroleum Structures*, Translated from the 1976 French edition by J. Chilton-Ward, Techniq, Paris, 1979a, 133.

Le Tirant, P., Seabed exploration by *in situ* measurements, in *Seabed Reconnaissance and Offshore Soil Mechanics for the Installation of Petroleum Structures*, Translated from the 1976 French edition by J. Chilton-Ward, Techniq, Paris, 1979b, 218.

Marlowe, J.I., A piston corer for use through small ice holes, *Deep Sea Res. Oceanogr. Abstr.*, 14, 129, 1967.

Martinais, J., L'instrumentation scientifique au Centre Oceanologique de Bretagne, *Colloque Inter. sur l'Exploit des Oceans*, Theme V, Vol. 1, 1971, 11.

Mawhinney, M.R. and Bisutti, C., *Common Corers and Grab Samplers Operating Manual*, Report, Technical Operations, National Water Research Institute, Environment Canada, Burlington, Ontario, 1987.

McIntyre, A.D., Deficiency of gravity corers for sampling meiobenthos and sediments, *Nature*, 231, 260, 1971a.

McIntyre, A.D., Efficiency of benthos sampling gear, in *International Biology Handbook No. 16*, Holme, N.A. and McIntyre, A.D., Eds., Blackwell Scientific, Oxford, 1971b, 140.

McMaster, R.L. and McClennen, C.E., A vibratory coring system for continental margin sediments, *J. Sed. Petrol.*, 43, 550, 1973.

Mellor, M., *Equipment for Making Access Holes Through Arctic Sea Ice*, Special Report 86-32, U.S. Army Corps of Engineers, Cold Regions Research and Engineering Laboratory, Hanover, New Hampshire, 1986.

Moore, T.C. and Heath, G.R., Sea-floor sampling techniques, in *Chemical Oceanography*, Vol. 7, 2nd ed., Riley, J.P. and Chester, R., Eds., Academic Press, New York, 1978, 75.

Mudroch, A., Sarazin, L., and Lomas, T., Report summary of surface and background concentrations of selected elements in the Great Lakes sediments, *J. Great Lakes Res.*, 14, 241, 1988.

Mudroch, A. and MacKnight, S.D., *Handbook of Techniques for Aquatic Sediment Sampling*, 2nd ed., Lewis Publishers, Chelsea, 1994.

Rosfelder, A.M. and Marshall, N.F., Obtaining large, undisturbed, and oriented samples in deep water, in *Marine Geotechnique*, Richards, A.F., Ed., University of Illinois Press, Urbana, 1967, 243.

Rowe, G.T. and Clifford, C.H., Modification of the Birge-Ekman box corer for use with scuba or deep submergence research vessel, *Limnol. Oceanogr.*, 18, 172, 1973.

Sly, P.G., Bottom sediment sampling, in *Proc. 12th Conf. Great Lakes Res.*, Inter. Assoc. Great Lakes Res., Ann Arbor, 1969, 883.

Sly, P.G., Equipment and techniques for offshore survey and site investigations, *Can. Geotech. J.*, 18, 230, 1981.

Sly, P.G. and Gardener, K., A vibro-corer and portable tripod-winch assembly for through ice sampling, in *Proc. 13th Conf. Great Lakes Res.*, Inter. Assoc. Great Lakes Res., Ann Arbor, 1970, 297.

Sundby, B., Silverberg, N., and Chesselet, R., Pathways of manganese in an open estuarine system, *Geochim. Cosmochim. Acta*, 45, 293, 1981.

Verrall, R.J. and Baade, D., *A Hot-Water Drill for Penetrating the Elsemere Island Ice Shelves*, Defence Research Establishment Pacific, DREP Technical Memorandum 82-9, 1982.

CHAPTER 4

Description of Equipment for Sediment Pore Water Sampling

4.1 INTRODUCTION

Sediment pore water, also referred to as interstitial water, is defined as the water filling the space between sediment particles and not held by surface forces, such as adsorption and capillarity, to sediment particles. The water content in typical sediments ranges from about 30%, for sand and mixtures of silt and sand, to up to 99%, for fine-grained surface sediments containing large amounts of organic matter. Sediment pore water acts as a linking agent between the bottom sediments and the overlying water. Negligible changes in sediment composition often cause noticeable variations in the quality of sediment pore water. The sediment pore water chemistry can help to explain many diagenetic processes. The increasing number of studies of contaminant concentrations in sediment pore water in marine and lacustrine environments reflect the significance of this part of the aquatic environment. However, the technique involved in collecting sediment pore water plays a crucial role in investigating pore water quality.

In studies of sediments, the sampling and analysis of sediment pore water can provide valuable information on chemical changes occurring in the sediment, on the equilibrium reactions between the sediments solid phase and water, on the transport and fluxes of contaminants into the sediment/water interface and overlying water, and on the availability of nutrients and toxic chemicals to the biota. The objectives of the study will define the appropriate method for sediment pore water sampling. Several techniques have been developed for sampling. Table 4-1 summarizes the most commonly used methods for sediment pore water sampling and their main advantages and disadvantages. There is no particular method for pore water sampling that can be considered ideal for all objectives or that is problem-free. The sampling requirements and methods of the techniques summarized in Table 4-1 are explained in more detail below. However, we emphasize that due to its com-

Table 4-1 Advantages and Disadvantages of Techniques Commonly Used for Pore Water Sampling

Method	Advantages	Disadvantages
Squeezing	Simple equipment Portable Inexpensive Sediment composition available	Oxygen contamination CO_2 degassing that will change pore water composition Temperature-induced changes Pressure-related additions of metabolites
Centrifugation	Simple Sediment composition available Easy to obtain large volumes	Risks of sampling artifacts Pressure artifacts Effects of oxidation and elevated temperatures
Dialysis	Minimal manipulation of sample No induction of interstitial water flow Allows maximal replication Analysis of dissolved gases is possible Temperature- and pressure-related artifacts are avoided	Disturbance of the sediment structure Need of scuba divers or submersibles Timing (minimum 12 days) Risks of incomplete equilibration Risks of membrane breakdown
Suction	Simple and easy to use Allows sampling at fairly well-defined depth	Fine particles could be collected (reduction of mesh size may cause clogging) Oxidation effects are hard to prevent

plexity, sediment pore water sampling is not suitable in all monitoring programs, and should only be carried out by personnel experienced with the sampling methods.

With the exception of the sediment/water interface, sediments are generally anoxic, and become rapidly oxidized upon exposure to air. The oxidation of the sediments induces immediate changes in the redox-sensitive chemical species of various dissolved elements in the pore water, usually with their subsequent precipitation. Therefore, the speed of the sampling procedure and the maintenance of an oxygen-free atmosphere are critical factors in sediment pore water sampling. The risks of contamination are considerable when sampling sediment pore water, due to the lower concentrations of elements in the pore water than those in the sediments. Therefore, to avoid contamination, the sampling and cleaning of all sampling equipment and containers for collection of the pore water need special attention.

Many factors, such as the objectives of the study, the requirements of special equipment, and time and cost constraints, should be considered before deciding on the sediment pore water sampling technique. In this chapter we present a descriptive overview of the different techniques and equipment available for sampling sediment pore water.

4.2 INDIRECT METHODS

When indirect methods of sampling sediment pore water are used, the pore water is recovered from sediments previously collected. The common advantage of indirect methods is the simultaneous collection of sediment, enabling comparison of the concentrations of analyzed parameters in the pore water to those in the sediment. The drawback of indirect methods is the need for maintaining the sediment under anoxic conditions to obtain a pore water sample representative of the sedimentary environment.

Considerable interest has been generated toward the role of sediment pore fluids in geochemical processes in deep marine environments. Among the topics of interest are the diagenetic reactions leading to the production of petroleum and natural gas; the origin of excess oceanic constituents such as boron; and the formation of authigenic phases such as manganese nodules or iron-rich deposits. The development of long core drill sequences has provided coring over a 1,000-m sediment depth. However, there are several distinct problems involved in the sampling of pore water from deep marine sediments. In deep oceanic waters it is very difficult to achieve the rapid sampling that could maintain the sediments in the laboratory in their *in situ* conditions. Consequently, there is an increased risk of changes in the geochemical character of the sediment induced by exposure to the atmosphere.

Usually, squeezing and centrifugation techniques are used to collect pore water in deep marine environments. Sediments are generally collected using piston cores (described in Chapter 3). Because the outer parts of the piston cores are frequently contaminated due to smearing of sediments along the core tube, Manheim (1976) recommended selecting the interior portions of the cores, which show the least sign of disturbance and deformation, for sampling pore water. With drill cores, the outer parts of cores may be more contaminated, because sea water is generally used as a circulating medium. Artificially compacted sediments, such as those in the butt of the core or in the coring head, should be avoided when sampling pore water. The differences in pressure and temperature from the sea floor to the surface inevitably affect the distribution of some chemical constituents in the pore water. The change in pressure affects the carbonate equilibria and may lead to exsolution of methane gas, which will result in the formation of gas pockets in the sediments (Manheim, 1976). Movement of the sea floor sediments from a temperature of 1 to 2°C to a room temperature of 20 to 22°C leads to changes in the distribution of cations in pore water (Mangelsdorf and Wilson, 1969; Fanning and Pilson, 1971).

4.2.1 Centrifugation

Centrifugation has been used for fluid removal, from various saturated or partly saturated geological materials, since the 1940s. Centrifuge extraction of sediment pore water for geochemical studies was established in the 1970s. The first methods had the following common problems:

- low speed of centrifugation, which was inefficient in separating the pore water from the sediments;
- long centrifugation time, which allowed physico-chemical changes and tended to heat the centrifuged samples; and
- contamination from the equipment.

Several groups addressed and tried to minimize these problems. Chemical artifacts from oxidation and temperature variations as a result of sample processing by centrifugation and filtration have been frequently described (Fanning and Pilson, 1971; Bray et al., 1973; Troup et al., 1974; Emerson, 1976; Lyons et al., 1979). Edmunds and Bath (1976) observed that low-speed centrifugation gave yields of 20 to 30% of pore water from the geological material of the Cretaceous Chalk aquifer in the United Kingdom, whereas at speeds of 14,000 rpm some 86 to 95% of pore fluid was obtained. In the same year, Emerson (1976) made considerable improvements in the centrifugation technique to study the early diagenesis in anaerobic lake sediments. The sediment fractions were placed in plastic sealed centrifuge tubes. The centrifugation was carried out at 5°C for 20 minutes at 20,000 rpm. The supernatant was then pressure-filtered through 0.45 μm filters using nitrogen gas. The first few milliliters that passed through the filters were discarded.

The use of syringes (with in-line filters) for withdrawing the supernatant pore water after centrifugation permits accurate and rapid transfer into containers for storage and analyses (Adams, 1994). As an example, Engler et al. (1977) placed sediments into oxygen-free polycarbonate centrifuge tubes in a glove box followed by centrifugation in a refrigerated centrifuge (4°C) at 9,000 rpm for 5 minutes. Extracted pore water (approximately 40% of the total) was vacuum-filtered under nitrogen through rinsed 0.45 μm membrane filters, immediately acidified to pH 1, and stored in rinsed plastic bottles. Adams et al. (1980) tested different centrifugation speeds from 7,000 to 19,000 rpm and found little change in pore water calcium, iron, manganese, and zinc, but phosphate doubled.

Elderfield et al. (1981) extracted sediment pore water from marine sediments in the laboratory, generally within 1 hour of sample collection. The sediment cores were sectioned inside a nitrogen-filled glove box at *in situ* temperature. The sediment (approximately 200 g) was transferred to gas-purged plastic centrifuge bottles, sealed, and centrifuged at 4,000 rpm for 20 minutes in a refrigerated centrifuge. The separated samples were filtered using back-to-back syringes separated by an in-line 0.4 μm Nucleopore filter. Aliquots of centrifuged but unfiltered pore water were also retained from some cores for analysis.

In 1985, Carignan et al. compared techniques for sediment pore water sampling. To evaluate the centrifugation and filtration method, sediment cores were sectioned into 1 cm slices in a glove box under a nitrogen atmosphere. Each section (about 40 cm^3) was transferred with a plastic spatula into a 50-ml polycarbonate centrifuge tube; the tubes were tightly closed and centrifuged at 5,000 rpm for 20 minutes. Following this, the supernatant was collected with plastic syringes and filtered through a Millipore 0.45 μm membrane using a Nucleopore in-line filter holder inside the glove box. The authors also tested centrifuging at 11,000 rpm and sequential filtration using Nucleopore membranes 0.2 and 0.03 μm. All final aliquots were stored for metal analysis at 4°C in clean polystyrene vials, preacidified with 50 μl of Ultrex 1N HNO_3. The investigators concluded that centrifugation at 5,000 rpm followed by filtration through 0.45 μm membranes was equivalent to dialysis for cobalt, nickel, chromium, iron, and manganese. However, it gave greater and more variable concentrations for copper, zinc, and organic carbon in the pore water. When the centrifugation speed was increased to 11,000 rpm and 0.2 and 0.03 μm membranes were used in the filtration, the concentrations of metals were comparable in both techniques.

To obtain pore water samples from the same depth intervals in marine sediments, Bauer et al. (1988) placed sediment sections in a simple pore water extractor consisting of a perforated plastic cartridge lined with a Nitex mesh insert and a glass-fiber filter. The entire assembly was transferred to a disposable plastic centrifuge tube (50 cm^3) and centrifuged at 1,000 × g for 5 to 10 minutes.

A simple pore water sampling device for simultaneous centrifugation and filtering of coarse, sandy sediments of low porosity (as low as 32%) was developed by Saager et al. (1990). The centrifuge tube (Figure 4-1) was made of high-density polyethylene, except for the snap-cap of linear polyethylene and the O-rings of silicon rubber or Teflon® for determination of metals in pore water. The centrifuge tube was closed with a tight-fitting snap cap to keep nitrogen gas inside, and centrifuging could proceed outside the glove box filled with nitrogen. Centrifugation was performed at 1,500 ×g for 5 minutes on board ship. After centrifugation, the lower part of the centrifuge tube was removed and the pore water obtained by pipetting the liquid. On average, 78% of available pore water was recovered, compared with 25 to 30% recovered by squeezing.

The elevated number of variables, mainly speed and duration, involved in the centrifugation of sediment for pore water extraction suggest that the measurements of parameters in collected pore water should be regarded as operationally defined (i.e., method-dependent) until it is shown that independent procedures consistently yield similar results (Carignan et al., 1985). However, there are a few practical considerations that may considerably improve the accuracy of the concentrations of parameters measured in pore water extracted by centrifugation of sediments. They include the following:

- minimize the risk of sample contamination as a result of the sediment coring process;

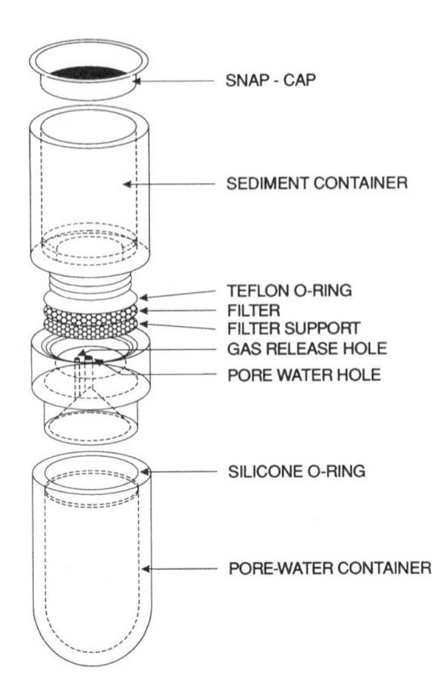

Figure 4-1 Sampler device for simultaneous centrifugation and filtering of sediments of low porosity (reprinted from Saager et al., *Liminology Oceanography*. 35(3), p. 748, 1990, with permission).

- reduce manipulation of sediments to avoid physical disturbances, such as always keeping sediment cores in a vertical position after retrieval;
- perform sediment subsampling and centrifugation in a refrigerated environment (4°C);
- perform all sample processing, such as subsampling, centrifugation, and filtration, in a glove box under an inert gas atmosphere.

4.2.2 Squeezing of Sediments

Sir John Murray, the pioneer British oceanographer, in conjunction with R. Irvine, was the first to squeeze fluids from a shallow Scottish coast sediment in 1895 (Manheim, 1976). Since then, modifications of filter presses have been used by petroleum workers for extraction of pore water from drilling mud. In the 1960s, the squeezing methods were adapted for extracting pore water from sediments (Siever, 1962; Hartmann, 1965; Reeburgh and Erickson, 1982). The designs were usually based on the principles of low- or high-pressure mechanical squeezing or low-pressure gas-mechanical squeezing, followed by filtration. However, all the first squeezers were made of stainless steel, with or without chrome plating, representing a potential source of contamination. In 1967, Presley et al. developed a new stainless-steel, low-pressure, gas-operated squeezer entirely lined with Teflon®, which ensured that the contact between the squeezed material and the metal surface was eliminated. The pore water was subsequently filtered through 0.4 μm Millipore filters. The authors were able to remove approximately 40% of the total water content by squeezing at a pressure of 70 kg/cm^2. In the same year, Reeburgh (1967) developed a non-metallic squeezer that was gas-operated and had no piston or moving parts (Figure 4-2). Delzyn and nylon were used for squeezer construction to prevent corrosion and for easy cleaning. Gas pressure (He, CO_2, and N_2) of up to 200 psi acted against a rubber diaphragm compressing the sediment and forced the pore water through filters into a sample bottle. Approximately 25 ml of pore water was extracted from 100 g of sediment in 30 to 45 minutes. Figure 4-3 shows a multiple clamping device for sediment squeezing designed by Rosa and Davis (1993). However, these gas-pressure squeezers necessitated relatively long periods of squeezing per sample in order to obtain a sufficient volume of water due to gas breakthrough and subsequent bypass.

Several studies (Bischoff et al., 1970; Fanning and Pilson, 1971; Froelich et al., 1979) also demonstrated that the temperature of squeezing played an important role in whether the sampled pore water was representative of the original one. Warming the marine sediment samples to room temperature prior to extracting the pore water accounted for up to 13% enrichment or depletion of some of the ions, such as potassium, calcium, chloride, and magnesium. (Bischoff et al., 1970). These differences were due to the changes in ion-exchange selectivity as a function of temperature. To overcome this problem,

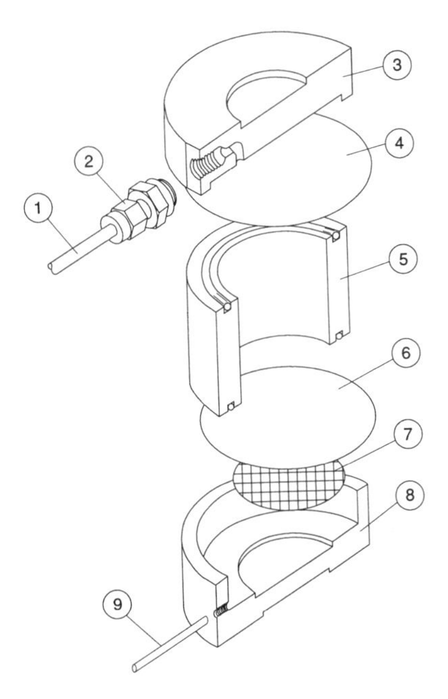

Figure 4-2 Exploded view of a gas-operated squeezer: (1) nylon gas inlet tube, (2) O-ring seal male plug, (3) Delrin cap, (4) dental dam rubber diaphragm, (5) nylon sample retainer with O-rings, (6) filter, (7) nylon screen, (8) Delrin base, and (9) nylon sample drain tube (modified after Reeburgh, 1967).

Figure 4-3 Multiple clamping device for sediment squeezing ("Quad-Clamp") (designed by Rosa and Davis, 1993).

Kalil and Goldhaber (1973) developed a thermoregulated squeezer system to operate at lake or ocean-bottom temperatures. In their design, the plastic liner in which the sediment was collected served as the squeezing chamber and was surrounded by an insulated water jacket. The model consisted of two identical plexiglass plungers with recessed, perforated Teflon® discs. Two filter papers were placed against the sediment surface. The whole assembly was mounted in a 3-ton press. Pressure was applied until water began to flow. The sample was collected when all the trapped air bubbles had been vented. Squeezing time varied from 10 minutes to 1 hour, depending upon the sediment characteristics.

Several groups of scientists have developed modifications to the squeezing technique in order to adapt this methodology to specific requirements in their study areas. For instance, Sasseville et al. (1974) created a large-volume sediment pore water squeezer for lake sediments. Large volumes of sample (as much as 800 cm^3) were placed in a PVC cylinder equipped with filter papers. The piston was inserted as far as the air hole, and the top was bolted on. Pressure was applied steadily by periodically turning down the piston screw. The pore water was collected in a polyethylene bottle fitted against the drainage hole of the squeezer. The water was immediately refiltered through a 0.45 μm Millipore filter.

Robbins and Gustinis (1976) modified the pistonless sediment squeezer developed by Reeburgh (1967) to be used on unconsolidated, fine-grained sediments. The squeezer was designed to work with small quantities of unconsolidated sediments (20 to 100 ml) from refrigerated cores sectioned in a nitrogen-filled glove box. To prepare the squeezers, all filters were purged with prepurified dry nitrogen and stored in a desiccator in a nitrogen atmosphere. After the prefilter was positioned in the cassette, about 30 ml of nitrogen-purged distilled water was added, and the system was purged with nitrogen (100 ml/min) for about 5 minutes. The time required to operate the extruder, section the sediment, and load the squeezer was about 1 minute. Cleaning the glove box and reassembling the squeezers took considerably longer, about 15 minutes per unit.

Matisoff et al. (1980) used nylon squeezers clamped to a squeezing rack for recovering pore water from the sediments of Lake Erie. Gas pressure (nitrogen at 3.4 atm) was applied against a rubber diaphragm to force the sediment pore water through two circles of nylon mesh supports overlain by two 0.45 μm Whatman filters and one 0.22 μm Millipore filter. Squeezing of sediment samples was carried out in a nitrogen-filled glove box to minimize exposure to oxygen.

In 1987, Bender et al. created a whole-core squeezer, to be used aboard ship, for pore water sampling with millimeter-depth resolution near the sediment-water interface (Figure 4-4). The squeezer consisted of a Lucite piston with a narrow hole through the center and an attached rod to which pressure was applied manually. Grooves 0.2 mm deep on the bottom of the piston channelled pore water toward the hole, where it was filtered through a 10 μm polyester screen. A lower piston, secured in the bottom of the core liner, retained the sediment. Collecting the water in 3-ml aliquots allowed sediment pore water profiles to be obtained with millimeter depth resolution. From the volume of the water squeezed, the radius of the core liner, and the porosity of the sediment, Bender et al. (1987) were able to calculate the depth of each pore water sample. The squeezing of the first two centimeters of sediment generally took from 30 to 60 minutes, but squeezing deeper, denser sediments was extremely slow. This method proved very useful for determining fluxes of O_2, NO_3^-, SiO_2, and other particle-unreactive ions and molecules in sediments. However, due to the rapid alterations by solid-solution reactions, this method is not suitable for determining pore water profiles of trace metals.

Bolliger et al. (1992) compared the extraction of sediment pore water by a squeezing technique with other methods. Sediment cores were sectioned immediately after retrieval under a nitrogen atmosphere into 0.5- and 1-cm-thick slices (i.e., 30 and 60 cm^3, respectively). Each individual sediment slice was squeezed in closed containers between a Delrin piston at the top and a PVC cap containing an outlet for collected water at the bottom. A 400 mesh nylon net (HD-30) and a glass-fiber filter (Whatman GF/C) were placed in the bottom cap as a prefilter to prevent clogging of the water outlet. Water samples were then collected using syringes in order to avoid air contact. The samples were subsequently filtered through a 0.45 μm membrane filter and distributed into different vials for analysis. The pressure, maximum 7,000 atm, was generated by a hydraulic system. Final pressures between 150 and 350 atm were necessary to obtain 10 to 30 ml of pore water from a 60 cm^3 sediment sample. The squeezing technique led to increased concentrations of Na^+, K^+, HPO_4^{2-}, H_4SiO_4, NO_2^-, and DOC. These increases were related to the destruction of either particulate detritus or sediment microorganisms under high pressure. It was suggested that pressure-related additions of intracellular metabolites

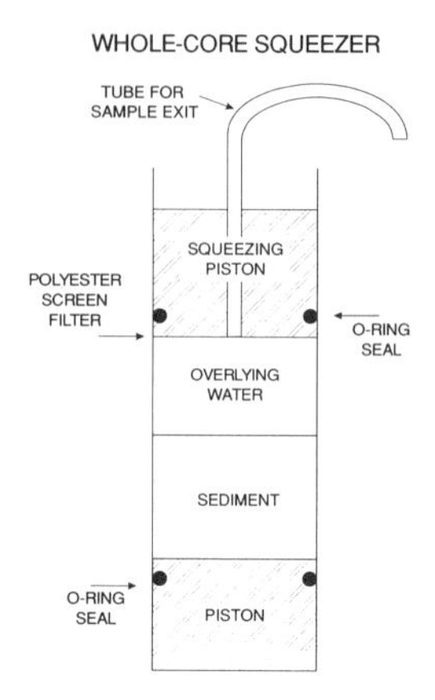

Figure 4-4 Schematic diagram of the whole-core squeezer (modified after Bender et al., 1987).

may be different in various sediment types and should be determined experimentally.

In summary, the major difficulties during the extraction of pore water by squeezing from anoxic sediments are oxygen contamination, CO_2 degassing, and release of different elements by destruction of sediment detritus and microorganisms. The oxygen contamination can oxidize dissolved elements, particularly iron, with subsequent precipitation and adsorption. Robbins and Gustinis (1976) showed that even traces of residual air on squeezer walls and in the filters could result in decreases up to an order of magnitude in concentrations of both dissolved iron and phosphate. Carbon dioxide degassing may result in calcium carbonate precipitation. Proper preparation of sediments before the actual squeezing plays an important role in maintaining the integrity of pore water chemistry (Adams, 1994). To avoid changes in the composition of the sediments collected for the extraction of pore water, sediment samples should be kept refrigerated until squeezing, with limited manipulations to minimize their disturbances. Because sediment bacteria are mechanically destroyed under the influence of high pressure, squeezing should be performed as gently as possible. We can conclude that this method is not suitable for quantitative extraction of pore water, but when employed properly it is a practical way to obtain representative samples of pore water for chemical analysis.

4.2.3 Gas Extraction and Vacuum Filtration Methods

Gas extraction methods involve passing gas of high relative humidity through water-saturated sediments to expel interstitial water. This effect is obtained by substituting the manual turn-screw of a squeezer for a gas-driven piston (Siever, 1962). Gas is introduced under pressure or is passed through the sediment by drawing it by vacuum. Released water drains through a discharge tube at the base of the press into a covered vessel to prevent evaporation (Scholl, 1963). These methods are not suitable for quantitative studies. The vacuum filtration technique is a very slow process and, consequently, evaporation of the pore water can be a critical factor when using this method. Gas extraction or vacuum filtration methods are not efficient for sandy materials of relatively low permeability, unless the driving pressures are higher than about 20 psi (1.4 kg/cm^2) (Scholl, 1963).

Using the same apparatus, extraction of the pore water can also be carried out by pouring an immiscible liquid over a sediment sample and forcing the liquid through the sediment. Ideally, the immiscible liquid should be less dense and considerably more viscous than the pore water in order to ensure a distinct fluid phase boundary. To keep evaporation to a minimum, the liquid should be chemically inert; it should also have low toxicity and price. A number of water-immiscible liquids have been investigated; of these, carbon tetrachloride, high-molecular-weight fluorocarbon, perchloroethylene, and trifluoroethane have the most desirable properties (Kinnlburgh and Miles, 1983). High-viscosity water-immiscible esters can be used for extraction in quantitative analyses of pore water. Yields of pore water from soils containing low (<49%) water content were typically 20 to 50% of the total water present (Kinnlburgh and Miles, 1983).

van Raaphorst and Brinkman (1984) developed a technique to collect pore water from undisturbed sediment cores under pressure. A 5-cm I.D. sampler was used to collect undisturbed sediment cores. The pore water was extracted with polyethylene tubing (2.5 mm I.D.) containing a thread of cotton. One of the ends of the tube was put into the core, and the opposite side of the tube ended in a 10-ml bottle with a rubber stopper. The pore water was collected creating underpressure to the sampling system by opening the valve on the bottle. Samples of about 5 ml were obtained within 2 or 3 days, depending on the sediment composition. This technique was employed to calculate transport coefficients and fluxes for phosphate and calcium across the sediment-water interface.

In 1985, Howes et al. developed a gas squeezer to collect pore water by pressurizing the headspace of a sediment core. The pressure was generated by syringe injection of argon or nitrogen, with direct transfer of water to a glass syringe as it flowed out through the glass tube. In 1988, Jahnke modified this sampler by adding top and bottom pistons for pressure application and tapped each hole to accommodate nylon screws and O-rings (Figure 4-5). Once the sediment core was in position the screws were removed and replaced with male-to-female luer fittings. Disposable filters and plastic syringes were then attached to each fitting. The top or bottom piston was pressurized with compressed gas (300 to 400 KPa) forcing pore water out of each fitting.

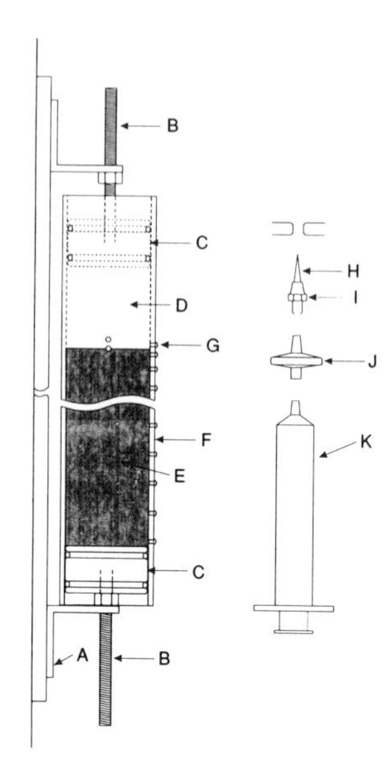

Figure 4-5 Schematic diagram of the pressurized core pore water sampler: (A) angle braces, (B) threaded rod, (C) piston seals with O-rings against the core barrel inner surface, (D) overlying water, (E) sediment core, (F) acrylic core barrel, (G) roundhead nylon screws from the depth intervals to be sampled, (H) threaded fitting, (I) female luer fittings, (J) disposable 0.4 μm filters, and (K) plastic syringes (modified after Jahnke, 1988).

4.3 *IN SITU* METHODS

Until 1970, pore water was extracted from sediments either by squeezing, centrifugation, leaching, or successive dilution. However, several investigators showed significant concentration changes mainly due to temperature- and pressure-related changes during the recovery of the pore water. The major problem during sediment pore water sampling and handling has been shown to be the potential for its oxidation. Therefore, many methods have been developed to collect pore water *in situ* to minimize these sampling artifacts.

4.3.1 Direct Suction

Since the early 1970s, several sampling systems have been proposed utilizing *in situ* sediment pore water suction and filtration. In 1973, Barnes designed an *in situ* sampler that filtered and encapsulated pore water from unconsolidated sediments. The sampler was designed to operate as an outrigger attached to a core barrel or mounted on a probe that was driven into the sediment. A variable delay time between tripping and initiation of sampling allowed the sampler to settle in the sediment. The pore water was filtered *in situ* under hydrostatic pressure through a three-layer stainless steel mesh filtering element, and collected in a stainless steel cylinder. Whiticar (1982) modified the

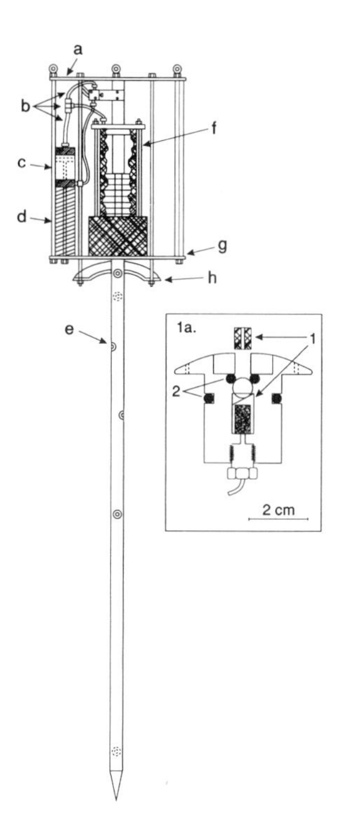

Figure 4-6 Schematic diagram of the sampler for *in situ* collection of pore water: (a) toggle valve, (b) lode pressure side tubing leading from master cylinder to valve and sampler interior, (c) master cylinder, (d) compressing spring, (e) port assembly in barrel wall, (f) "pot" containing slave cylinders and capillary spools, (g) base plate, (h) tripping plate. 1a is a detail of the one-way port assembly: (1) PVC spacers, (2) O-ring seals (reprinted from Sayles et al., 1976, 261, with permission from Pergamon Press, Ltd., Oxford).

sampler designed by Barnes to collect *in situ* pore water and gases. The modified sampler included replacement of the delay valve mechanism, which prevented premature filtering, with a stainless steel rupture disk punctured approximately 10 seconds after the sampler penetrated the sediment.

Sayles et al. (1973) developed a sampler for *in situ* collection of pore water from marine sediments. The instrument (Figure 4-6) consisted of a 2-m long stainless steel tubing (replaced in 1976 by a thick-walled aluminum tubing) with a pointed tip for penetration and five filter-covered sampling ports (6 cm^2 each). A broad base plate was placed above the sampling ports to prevent penetration. One port above the base plate was used to sample overlying water. The instrument functioned as a large syringe with a spring-loaded cylinder providing the suction. Before lowering the probe into the sediment, the cylinder was charged hydraulically to a pressure of 34 atm. Once the instrument penetrated the sediments (up to 2 m), a tripping ring triggered a toggle valve, releasing the compression and initiating the sampling. Pore water was drawn with syringes from the six sampling ports, filtered, and stored in capillary tubing. After about 30 minutes of sampling, the device was retrieved. Sayles (1979) pointed out two problems with this sampler. Over-penetration of the sampler into the sediments could result in an incorrect assigned depth, and disturbance of the probe during the sampling period could cause leakage along the barrel and contamination of pore water with overlying bottom water. Murray and Grundmanis (1980) replaced the Teflon® capillary system with nylon sample loops for pore water gas analysis.

Porous ceramic samplers and Teflon® samplers, commonly used for determination of nutrients in soil solutions, were adapted to measure *in situ* nutrients in sediments (Zimmermann et al., 1978). A porous cup attached to a length of PVC pipe was inserted at the desired depth in the sediment (Figure 4-7). The sampler was allowed to equilibrate for 48 hours. After 48 hours, vacuum was applied through a vacuum/purge line into the sampler. The pore water entering the ceramic cup during a predetermined time was collected through the line. Recoveries of pore water using ceramic cups ranged from 11 to 111%, while Teflon® samplers consistently recovered 98 to 106%. The authors indicated that 125 ml of pore water could be collected during one day over a fourteen-day period without clogging the Teflon® sampler or reducing the sample volume. This method had two main disadvantages: it required one sampler for each sediment depth and produced large variations due to heterogeneity of the sediment.

To overcome these limitations, the same group developed a more complex sampler designed to sample the pore water simultaneously and continuously from four sediment depths (Montgomery et al., 1981). The Plexiglass sampler was composed of four independent chambers with 1-cm porous Teflon® filters each (Figure 4-8). A top section provided the necessary connections to the internal sampling chambers. Each chamber had two tube connectors: a sampling tube, and a tube for the application of vacuum or purge gas. Once inserted into the sediments, the sampler was purged with freon gas, sealed, and equilibrated for 72 hours. After 72 hours, pore water was sampled using a vacuum pump at a flow rate of 4 ml/min; the first few milliliters were discarded.

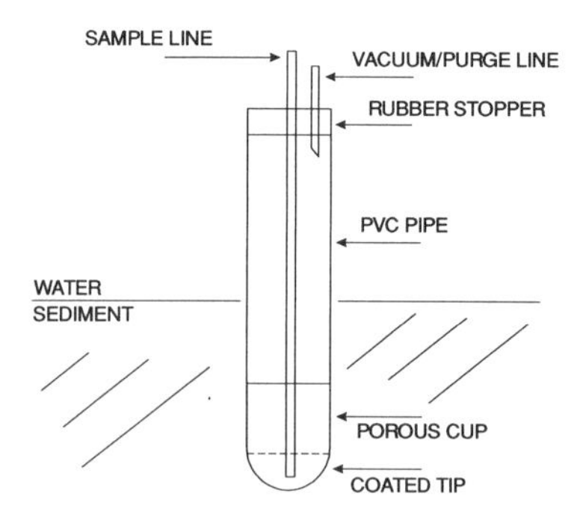

Figure 4-7 Porous sampler (reprinted from Zimmermann et al., *Estuarine Coast Mar. Sci.*, 1978, with permission).

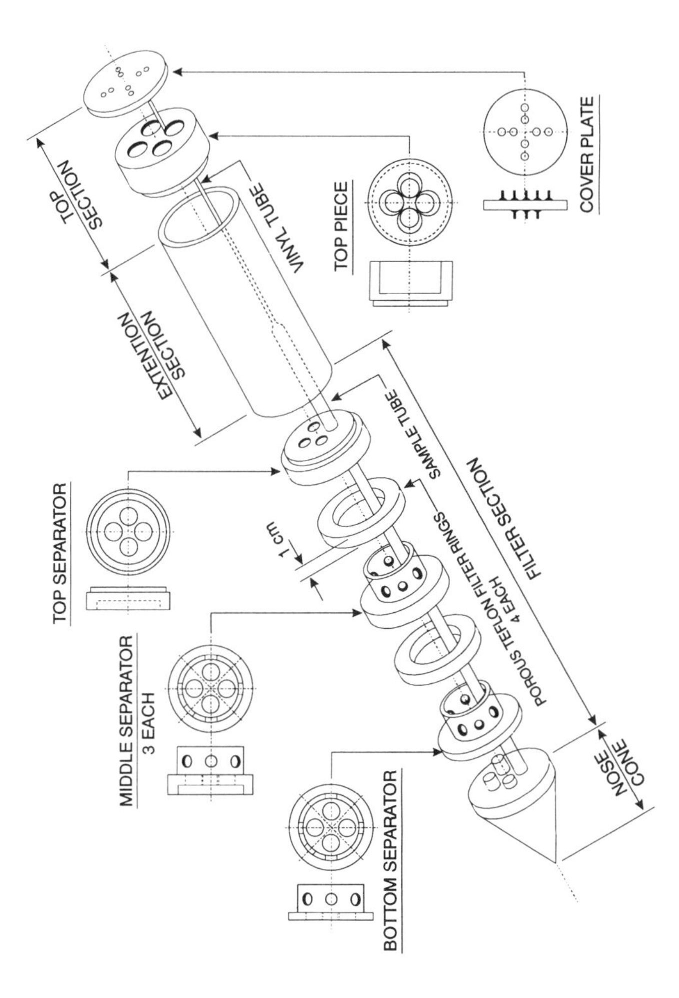

Figure 4-8 Exploded view of a sampler for collection of pore water simultaneously and continuously from four different depths (after Montgomery et al., *Estuaries*, 4, 76, 1981, with permission).

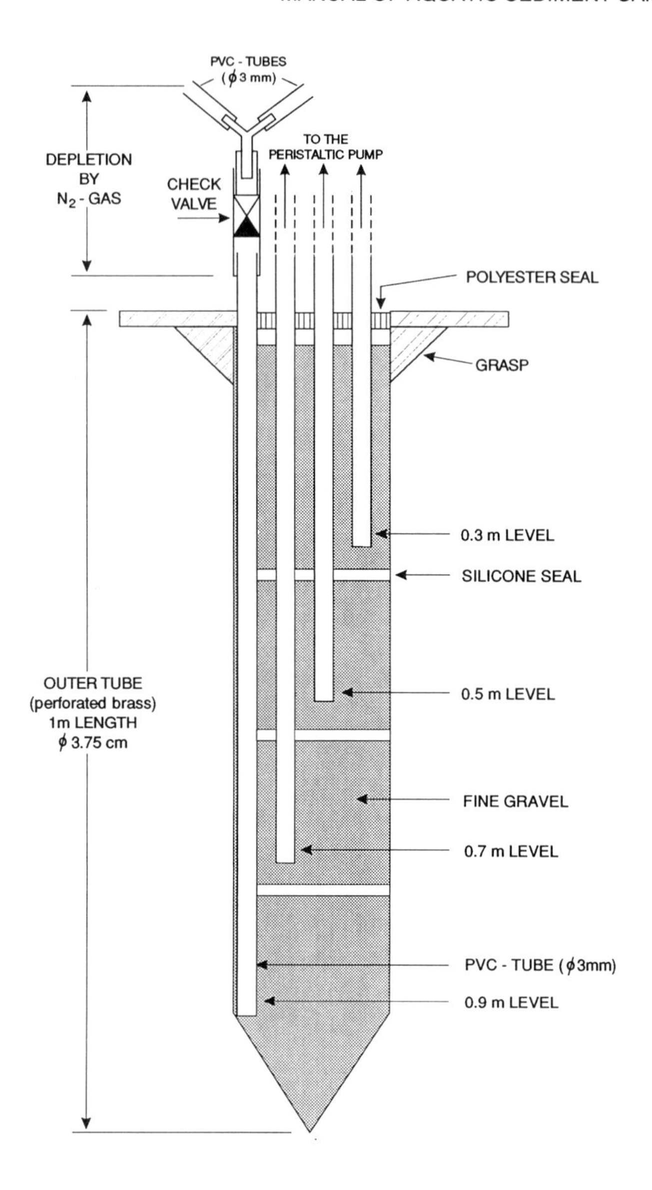

Figure 4-9 Cross-sectional view of a sampler designed to collect pore water from rivers and lakes at different depths (after Hertkorn-Obst et al., *Envir. Tech.*, 3, 264, 1982, with permission).

A similar sampler (Figure 4-9) was designed by Hertkorn-Obst et al. (1982). Four or more PVC tubes were used at 10 to 20 cm depth intervals, depending on the final depth reached in the bed. The withdrawal of pore water at each level was achieved by connecting the respective tube to a slow peristaltic pump and extracting water at a rate of about 400 ml per hour. Recently, Rey et al. (1992) designed a modified version (Figure 4-10) of the sampler de-

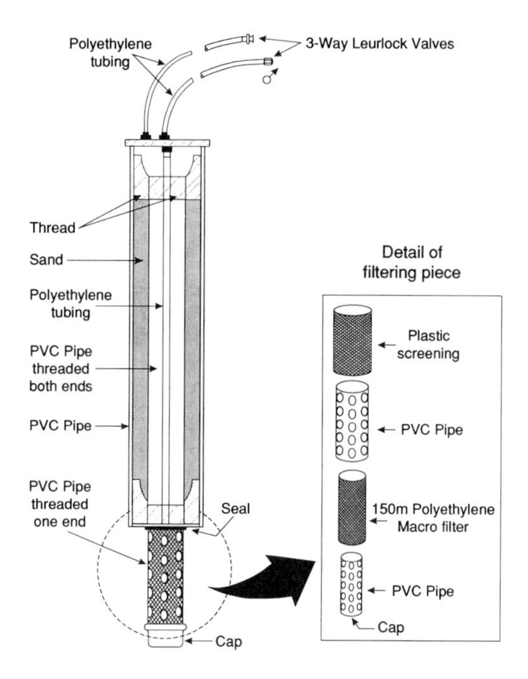

Figure 4-10 Schematic diagram of an *in situ* pore water sampler with filtering unit (modified after Rey et al., 1992).

scribed by Zimmermann et al. (1978). The filtering unit of this sampler had four pieces: (1) an external plastic screening, (2) a PVC pipe, (3) a 150-μm polyethylene macro filter, and (4) an internal PVC pipe. However, this new design has the same limitations as the original sampler; it requires one sampler for each sediment depth, and is limited by heterogeneity of the sediments.

Goodman (1979) described a very simple instrument to directly collect pore water from sandy sediments. It consisted of a probe with a 3-mm diameter and 0.45-μm mesh that was introduced into the sediments. A capillary tube of 1 mm I.D. connected this probe with a syringe for collection of the pore water sample. Based on the same principle, Brinkman et al. (1982) designed a sampler that allowed the collection of pore water from defined sediment depths. Up to ten probes could be attached to a stainless steel frame, which was pushed into the sediment with a bar (Figure 4-11). Metal plates on the legs of the frame prevented the sampler from sinking into the mud. The probe consisted of two 10-mm diameter tubes (Figure 4-12). The length of the second tube was variable, depending on the desired sampling depth. Paper or glass-fiber filters were placed between both tubes. The pore water sample filtered horizontally through the filter, and three vertical groves in the screwthread allowed the pore water to enter the tube. The screw that connected these tubes with a manual valve contained a float. When the sampler filled, the float prevented overflow of the sampler with the pore water. The other probes that were connected to the vacuum lines of the frame were not affected. Sampling time was approximately one day.

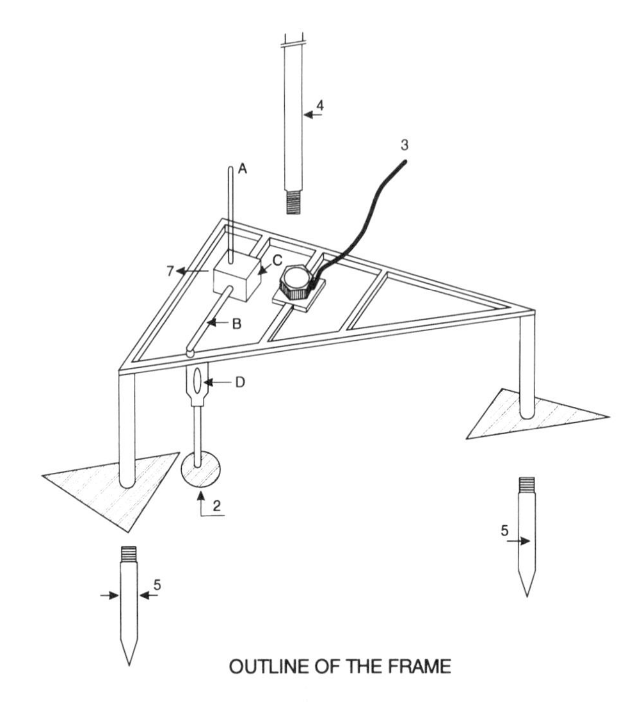

Figure 4-11 Outline of the frame of an *in situ* pore water sampler for shallow lakes: (1) metal plate, (2) covering disk, (3) recovery rope, (4) pushing bar, (5) under part of legs (can be screwed off), (6) central magnetic three-way valve. (A) tube to nitrogen source, (B) frame-vacuum line, (C) outlet of magnetic three-way valve, (D) outlet of manual three-way valve (after Brinkman et al., *Hydrobiologia,* 92, 660, 1982, with permission).

A sampler ("dipstick") for continuous sulphide profiles in marine sediments was described by Reeburgh and Erickson (1982). A thin (2 mm) band of polyacrylamide gel was cast in a dipstick sampler or Lucite support. The gel was doused with specific color-forming reagents that reacted with components in the sediments and provided a visual semiquantative indication of depth distributions. Reeburgh and Erickson used gels that were doused with lead acetate for studying the distribution of sulfide in anoxic marine sediments. Since the gels are 90% water, reaction with pore water is controlled by diffusive transport. Up to 90% equilibration in sediments was obtained within 30 minutes. The sulfide concentration in the pore water was calculated based on the degree of darkness in the gel, the exposure time, and the lead concentration in the gel (new gel samples described in Section 4.3.3).

Howes et al. (1985) developed a sampler consisting of glass capillary tubing with a perforated Teflon® sleeve sealed at the bottom with a glass plug (Figure 4-13). A serum stopper was attached to the top for removal of pore water. The samplers were left in the field for at least one week before the first sampling and subsequently sampled for weeks to months. All samples (5 to 10 ml) were collected with glass syringes. The main disadvantage of this tech-

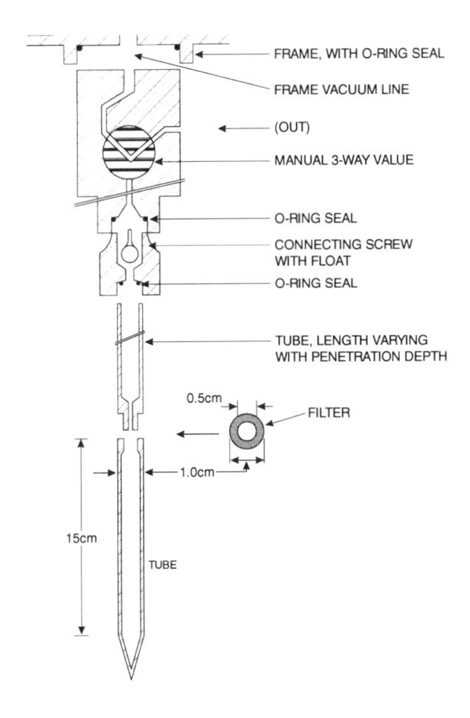

Figure 4-12 Schematic diagram of the probe of the *in situ* pore water sampler for shallow lakes (after Brinkman et al., *Hydrobiologia,* 92, 660, 1982, with permission).

nique is the uncertainty about the precise depth interval measured, since water must flow to the sampler during collection.

Pore water was collected by scuba divers using direct insertion of syringe needles (13 gauge) into the sediments (Bauer et al., 1988). Long needles were marked at 0.5-cm intervals and attached to plastic syringes. Approximately 1 to 5 ml of pore water was collected into the syringe from each depth by the diver.

A sampler for *in situ* separation of pore fluids from intertidal sediments and laboratory microcosms tests has been built by Watson and Frickers (1990). The sampler (Figure 4-14) is operated by applying a vacuum to chambers linked, through porous segments, to successive depths in sediment profiles. The basic module consists of a solid acrylic cylinder (6 cm I.D., 10 cm long) with a lower 4-cm section machined to a 70° taper. A series of five holes (2 cm I.D.) were drilled vertically into the cylinder. The open end of each sample chamber was fitted with a threaded insert into which an acrylic cap with an O-ring seal was screwed. After the sampler was inserted into the sediment, the free ends of the withdrawal tubes (Figure 4-14) were attached with luer connectors to one-way Teflon® valves fitted to an anaerobic jar containing individual sample collection tubes. In consolidated and sandy sediments it was necessary to first remove a subcore to aid penetration of the sampler. To minimize the effects caused by the introduction of oxygen or oxygenated surface water into the

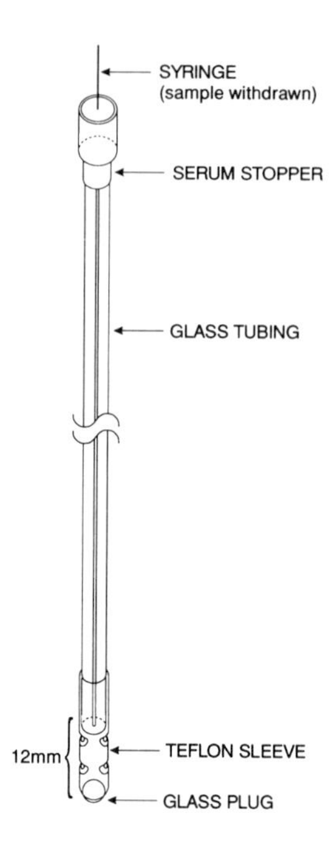

Figure 4-13 Schematic representation of an *in situ* sampler to collect pore water by pressurizing the headspace over the core by syringe injection of argon or nitrogen (modified after Howes et al., 1985).

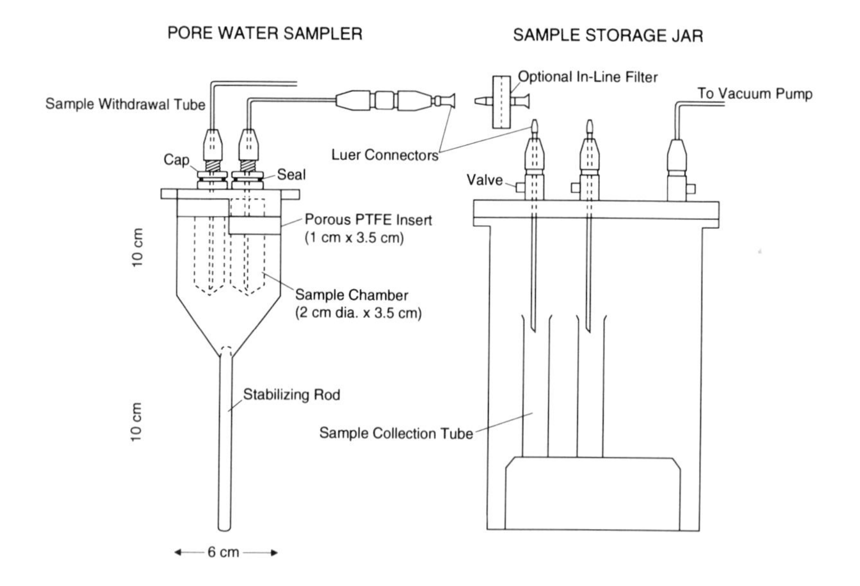

Figure 4-14 Schematic diagram of the multileveled *in situ* pore water sampler and sample storage vessel (modified after Watson and Frickers, 1990).

sediments, the authors left the sampler for 24 hours to equilibrate. The elevated machine costs were one of the limitations of this technique.

Recently, Hursthouse et al. (1993) developed a similar *in situ* sampler (Figure 4-15). It consisted of a polyethylene tube with 1.5-cm diameter sampling ports drilled in a spiral sequence 2.5 cm apart and stoppered with rubber bungs. The tube acted as a "template" and was carefully inserted in a precored hole in the sediment. The filters were inserted approximately 5 cm into the sediments to reduce the risk of contamination by surface water percolating down the sides of the template. Sediment pore water was collected by connecting the free ends of the tubing to a nitrogen-flashed sampling chamber (Figure 4-14) and a vacuum was applied using a simple hand pump.

4.3.2 Dialysis

Samplers based on diffusion-controlled transport were first developed by Mayer (1976) and Hesslein (1976). The principle of operation of these samplers is the equilibration between water contained in the sampler and sediment pore water through a dialysis membrane. The technique from Mayer consisted of placing dialysis bags filled with particle-free distilled water into the sediment for a period of time sufficient for equilibration with the sediment pore water. After the equilibration, the dialysis bags were retrieved, cut open, and the pore water was collected in a sample bottle. For profiles of pore water, Mayer used a sampling device consisting of a perforated Lucite tube separated into chambers containing individual dialysis bags. The device was pushed into the sediment, leaving one or two chambers exposed above the sediment/water interface. Hesslein's sampler, also called a "peeper" or dialyzer sampler, consisted of individual compartments machined into two sheets of

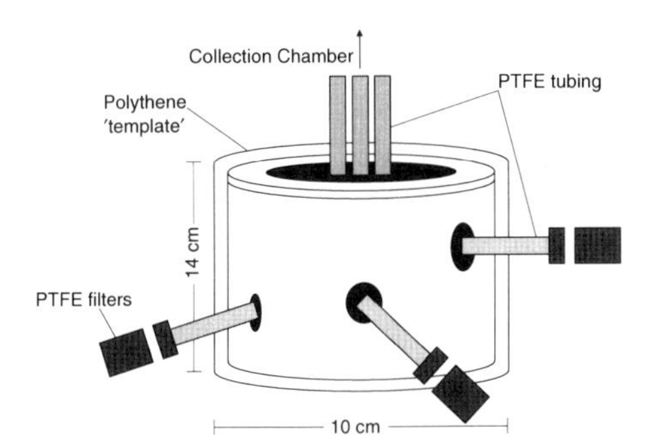

Figure 4-15 Schematic diagram of an *in situ* pore water sampler to be inserted into a precored hole in the sediment (modified after Hursthouse et al., 1993).

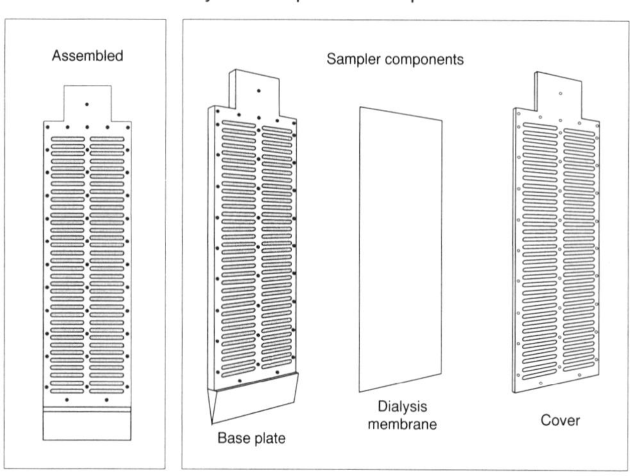

Figure 4-16 Front view and components of a dialyzer sampler ("peeper").

acrylic plastic with a dialysis membrane placed between the sheets (Figure 4-16).

Höpner (1981) developed a diffusion sampler for fine-grained aerobic sediments. The sampler consisted of a stainless steel rod 50 cm long, with twenty chambers (3 ml each) over a distance of 33 cm. The chambers were covered by a membrane that was mechanically protected and held in place by a stainless steel plate. The lower end of the sampler was pointed and the upper end fitted to a board to facilitate insertion into the sediment. A few modifications were made to the sampler for collection of pore water in anoxic sediments. A plastic tube was driven into the sediment around the sampler so that an external core of sediments was present during retrieval.

Bottomley and Bayly (1984) modified the dialysis bag technique. The sampler or "cylindrical dialysis probe" consisted of a perforated Lexan sleeve (40 cm × 3 cm) with a solid tapered base and a threaded top portion that was capped (Figure 4-17). The probe was compartmentalized by stacking ten small vials. Each vial had ports covered with dialysis membranes that were glued and secured with O-rings. The vials were filled with deionized water and deoxygenated by overnight equilibration in a closed bath of distilled water through which nitrogen gas had been bubbled. Each vial held a 10- to 12-ml sample, which was emptied by hypodermic syringe through the rubber septa positioned at the ends.

Membrane Dialysis — "Peeper"

The peeper sampler developed by Hesslein (1976) and its modifications are one of the most common sampling techniques for *in situ* sediment pore water

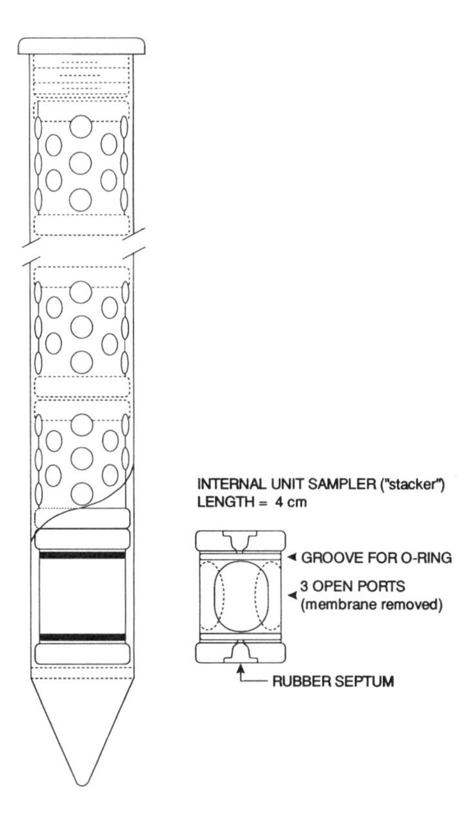

Figure 4-17 Schematic diagram of the cylindrical dialysis *in situ* sampler ("stacker") (modified after Bottomley and Bayly, 1984).

sampling. Therefore, this manual will deal in more detail with the preparation and retrieval of peepers. A modified version of Hesslein's sampler is shown in Figure 4-16. Generally, the samplers are made of clear acrylic plastic, such as Plexiglass or Lucite. The original sampler developed by Hesslein in 1976 consisted of two sheets of acrylic: one 0.3-cm cover and a 1.3-cm-thick body, held together by a series of nylon or stainless steel screws. Elongated sampling compartments are machined in the acrylic body, usually 1 cm apart. The 0.3-cm cover has slots cut in the same positions as the compartments in the body.

Although the principle of this *in situ* pore water sampler has remained the same, many scientists have modified the design of the sampler to suit their specific needs. For example, some samplers have larger compartments for studies where larger volumes of pore water are necessary, or the peepers may have thinner walls to facilitate penetration in compacted sediments. The use of dialysis or other membranes allows for discrimination of particles or molecules of different sizes entering the sampler chamber, and can eliminate filtering of the sample. Simon et al. (1985) modified Hesslein's sampler designing two 3-mm-thick cover sheets and 0.2 μm polycarbonate membranes mounted in the sampler with the glossy side out to provide less surface area for attachment of different microorganisms. Because of microbial attack, cellulose-based dialysis membranes are not recommended. Carignan et al. (1985) observed that a reduction of the dialysis membrane pore size of 0.03 μm to 0.001 μm indicated either incomplete equilibration after two weeks, or exclusion of medium sized metal-organic complexes. Modifications of the Hesslein sampler were also described by van Eck and Smits (1986). The new sampler was called a memocell. The major modifications were the use of a second solid acrylic plate and the construction of side ports for collecting the sediment pore water. Pore water was withdrawn into special 10-ml plastic monovette syringes (Sarstedt, Numbrecht-Rommelsdorf, Germany), complete with septor, through a sliding apparatus mounted over the side ports of the sampler. Kepkay et al. (1981) designed another pore water sampler with shutters positioned over the dialysis membrane.

Figure 4-18a shows a diver inserting the peeper vertically in the bottom sediments of a lake. Figure 4-18b shows the peeper after deployment in the sediment with approximately 16 cm of the upper part of the peeper above the water-sediment interface. Figure 4-19 shows the collection of the pore water samples through the membrane after retrieval of the peeper from the bottom sediments. An extensive review of applications of membranes of different pore sizes in sediment pore water sampling was provided by Adams (1994).

Procedure for Peeper Preparation and Deployment

A list of the equipment required for peeper assembly is summarized in Table 4-2. To avoid cross-contamination, cleaning of the peepers is the most important stage in preparation before assembly. Peeper cells are cleaned individually with a small brush before placing the peeper in a bath containing 1M HNO_3 for two weeks. It is recommended to leave new peepers in the bath for four weeks. The peeper is prepared for sampling by filling the compartments with deoxygenated double-distilled water (DDW) with subsequent careful placement of a piece of dialysis membrane over all compartments (Table 4-3). However, it is essential that all air bubbles are removed from the peeper cells during assembly. Once the peeper is assembled, it is placed in a portable polyethylene tank, or bubbling chamber, previously filled with deoxygenated DDW. The bubbling chamber with the assembled peepers is purged with nitrogen-gas for at least 24 hours prior to the placement of the peeper into the sediment to maintain anoxic conditions. The peepers are transported to the sampling site in sealed bubbling chambers containing nitrogen-purged water.

At the sampling site, the peepers are removed from the bubbling chambers and inserted vertically into the sediments, preferably in areas with fine-grained sediments and flat bottom surfaces. In areas with prevalent waves or tides, the sampler is oriented with the narrow side parallel to the direction of the waves (or tide) and manually pressed into the sediment. This orientation minimizes the dislocation of the peeper by currents during the sampling period. When

a

b

Figure 4-18 (a) Scuba diver inserting an *in situ* dialyzer (peeper). (b) Sampler after deployment in the sediment.

visibility permits, divers should record the number of peeper cells above the sediment-water interface after placing the peeper into the sediment to estimate sediment disturbances and movement of the peeper during the sampling period. Apparatus called "peeper placer" (Savile and Pedrosa, 1980) or "aluminum benthic landers" (Mueller and McNee, 1994) can also be used if scuba diver support is not feasible or in water bodies where water depth exceeds 25 m (Figure 4-20). Peepers placed in the sediment are attached by a rope to an anchor. The water in the peeper chambers must be allowed to equilibrate with the sediment interstitial water. In different studies, the reported time necessary

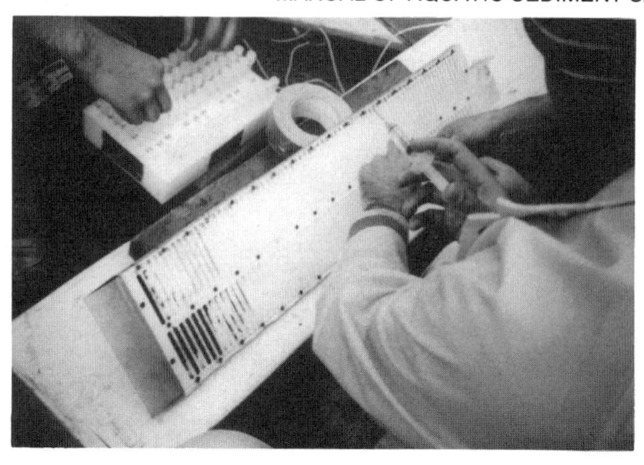

Figure 4-19 Sampling pore water through the membrane of an *in situ* dialyzer.

Table 4-2 Equipment Required for Peeper Assembly and Retrieval

(a) Assembly
- Doubly distilled water
- Deoxygenated water
- Nitrogen gas cylinder with regulators
- Polyethylene containers (washing tank, assembling tray, and water storing)
- Dialysis membrane
- Peepers
- Cleaning utensils: brushes, glass rod, and plastic bar
- Screws and screwdriver (battery operated recommended)
- Rubber tubing for releasing bubbles
- Bubbling chamber

(b) Retrieval
- Peeper board (to support peepers during emptying)
- Syringe and needles, boxes for syringe disposal
- Sample containers (polypropylene or scintillation vials)
- Pipettes
- Numbered racks or boxes of scintillation vials to hold sample containers during sampling in the boat
- Preservative solution (HCl, H_2SO_4, etc.)
- Tape
- Coolers
- Water-proof pen
- Ropes
- Field notebook

for equilibration varied between six and thirty days. The most important factors affecting the equilibration time are the diffusion coefficient of the substance of interest, its adsorption to the solid phase, temperature, and porosity of the sediments (Carignan, 1984). Generally, for most recent sediments with temperatures such as 4–6°C and 20–25°C, twenty and fifteen days, respectively, appear to be safe equilibration periods for major ions and nutrients (Carignan, 1984).

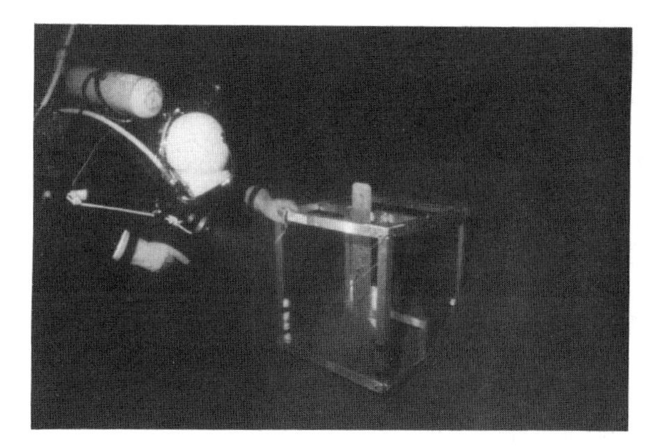

Figure 4-20 Aluminum benthic lander used to place peeper in deep lake waters.

Table 4-3 Procedures for Cleaning and Assembly of Peepers

1. After cleaning interior of individual cells with a small brush, place peeper in an acid bath (1M HNO_3) for two weeks.
2. Keep nitogen bubbling at slow rate for two days in a clean plastic tank, with lid, filled with double-distilled water (DDW).
3. Place peepers and peeper cover plates that have been acid-soaked and rinsed with DDW into a large polyethylene tray (approximately 55 cm × 80 cm ×14 cm) filled with deoxygenated DDW.
4. Remove ALL air bubbles from the peeper cells with acid-cleaned utensils such as a plastic-bristled brush, glass rods, and plastic bar.
5. Submerge dialysis membrane (approximately 3 cm longer than peeper length) in the tray in between the peeper and the peeper cover. Make sure they are bubble-free.
6. Once the peeper, cover plate, and membrane are lined up, place screws in the corners and every fourth side screw. Take special care not to perforate the cell membrane.
7. Inspect the peeper for air-bubbles. Once all the bubbles are removed, put in all screws to make the peeper airtight.
8. Remove all assembled peepers from the tray and place in a portable polyethylene tank (bubbling chamber) filled with deoxygenated DDW. Maintain in bubbling nitrogen until its transportation to the sampling site.

Procedure for Peeper Retrieval

Table 4-2 lists the principal equipment required for retrieval of the peepers. As already described, equilibration times of fifteen to twenty days were found adequate for most recent sediments. Shortly before the peepers are retrieved, the appropriate type and quantity of preservative, based on the analysis to be performed, should be added to all previously washed sample containers, including a few spare ones. The cleaning and preservative addition requirements for different analyses of sediment pore water are discussed in Section 6.5. All acidified containers should be placed in trays for manipulation on the sampling boat or on the platform where the samples will be recovered from the peepers.

The trays should provide enough space between sample containers for easy uncapping and capping, and clearly show the sample number. Before retrieval of the peeper, two boxes should be prepared, one containing new syringes with an attached hypodermic needle, and a second for the disposal of the used syringes. One syringe per cell must be used to avoid cross-contamination of the samples.

It is recommended to complete the sampling of the pore water from all chambers in the peeper within 5 minutes of retrieval from the sediments. A delayed sampling increases the risk of oxidation of the pore water during the sampling procedure, which could significantly change the concentrations of dissolved elements, such as phosphorus and iron. It is therefore essential to efficiently and quickly sample the pore water from a peeper retrieved from anoxic sediments. The peepers are removed from the sediment by a diver or pulled by a rope from the boat. Once at the water surface, the peeper is shortly agitated in the water to remove any visible sediment particles adhering to the membrane surface. The compartments are sampled by tilting the peeper from a horizontal position to facilitate drainage of the cells (Figure 4-19). The pore water is withdrawn from the lower end of the cell by syringe. Special care must be taken to avoid drawing air into the syringe to prevent oxidation of the sample. It is highly recommended to have three or four people for the sampling of the pore water from the peeper: two people to empty the peeper cells using the syringes, and one or two to empty the syringes into the appropriate numbered and pretreated sample containers at the opposite side of the peeper. A record for each peeper retrieved should include the following: the visual sediment-water interface observed by the diver; the position of the oxic-anoxic interface as indicated by the precipitation of reddish-brown ferric oxyhydroxide staining on the peeper membrane; and the time between the removal of the peeper from the sediment and emptying the last cell. More details regarding the preparation of peepers may be found in a report recently prepared by Rosa and Azcue (1993).

4.3.3 Ion-Exchange Resin with Gel Membrane

Dialysis with receiving resins has been used for determination of trace elements in seawater (Davey and Soper, 1975; Reeburgh and Erickson, 1982) and polluted waters (Morrison, 1987). Davison et al. (1991) developed a new technique of diffusive equilibration in a thin film to study the distribution of dissolved iron in sediment pore water at submillimeter resolution. This technique relies on the equilibration principle, as does the dialysis techniques, but rather than confining the solution to compartments, it uses a thin film of gel to provide the medium for solution equilibration. The main advantage of this technique is the short time for solution equilibration. Using a 1-mm-thick gel and assuming a diffusion coefficient of 10^{-5} cm^2/s, complete equilibration takes place in 42 minutes (Davison et al., 1991).

The sampler used by Davison and colleagues consisted of a polyacrylamide gel cast between two glass plates separated by plastic spacers to give final dimensions of 3.1 cm ×15 cm × 0.1 cm. To remove the oxygen, the gel was equilibrated for 24 hours in water bubbled with carbon dioxide. The gel system was then immediately inserted into the sediment until the front window was half exposed to the sediment and half to the overlying water. After equilibration for 24 hours, it was removed from the sediment and rinsed with distilled water before immersion for 2 hours in a solution of 1 mmol/l sodium hydroxide to convert Fe(II) into insoluble Fe(III). After this treatment, the gel was washed, peeled from its backing plate, and then pressed between either Teflon® or acrylic sheets before drying at 60°C. The gel shrank on drying, and the new thickness of 50 to 100 μm was measured by scanning electron microscopy to provide a concentration factor for each film. Iron concentration in the dried gel was determined using PIXE. Concentrations in the pore waters were calculated from the measured volumetric concentration factor and the density of the gel, assuming that 90% of the wet gel consisted of water capable of equilibration with the pore waters. Davison and colleagues recommended a general fixation procedure, such as freeze-drying, to determine diffusing components other than iron.

Recently, Davison and Zhang (1994) described a new technique for determining speciation of trace components in pore waters. The technique incorporates an ion-exchange resin separated from the solution by an ion-permeable gel membrane. Transport in the gel is restricted to diffusion and, by selection of an appropriate gel thickness, it controls the overall rate of mass transport irrespective of the hydrodynamics in the bulk solution. Providing the resin is not saturated, the longer the sampler is immersed in the sediments, the more metal will be accumulated, and the ratio of the concentration of metal in the resin layer to metal in solution will increase as the thickness of the resin and gel layers are decreased. For 24 hours immersion of a gel layer 1 mm thick and a resin layer 0.1 mm thick, and assuming a diffusion coefficient of 10^{-5} cm^2/s, the concentration in the resin layer will be 864 times greater than the concentration in the pore water solution (Davison and Zhang, 1994). This method of a resin backing a gel relies on establishing a defined diffusion gradient, in contrast to the previous one (Davison et al., 1991), which depended on the equilibrium being established.

The samplers used by Davison and Zhang (1994) consisted of polyacrylamide gel, backed with a thin film (≈150 μm thick) of gel containing a cation exchange resin selective for trace metals (Chelex 100) close-packed in a single layer of 75 to 100 μm spheres. The gel was exposed through a 5-cm diameter window; a set of screws and an O-ring prevented water from being introduced to the side or back of the gel. After casting, polyacrylamide gel was hydrated in water for at least 24 hours. The resin was embedded in a separate ≈150-μm-thick gel as a single plane of approximately close-packed beads. After removal from the sediment, the gel layer was peeled off, metal extracted from the resin layer with 1 ml of 2M HNO_3, and measured using atomic absorption spectrophotometry (AAS). The thickness of the hydrated gel was measured by a travelling microscope and was accurate to better than 4% (Davison and Zhang, 1994). This new technique is suitable to study dynamic situations, such as the variation in sediment-water fluxes during river flooding, or temporal changes in seawater. However, it requires further testing in environments such as fresh waters, where colloidal material such as fulvic complexes could adsorb directly to the resin.

REFERENCES

Adams, D.D., Sediment pore water sampling, in *Handbook of Techniques for Aquatic Sediments Sampling*, 2nd ed., Mudroch, A. and MacKnight, S.D., Eds., Lewis Publishers, Chelsea, Michigan, 1994.

Adams, D.D., Darby, D.A., and Young, R.J., Selected analytical techniques for characterizing the metal chemistry and geology of fine-grained sediments and interstitial water, in *Contaminants and Sediments*, Vol. 2, Baker, R.A., Ed., Ann Arbor Science Publisher, M.J., 1980, 3.

Barnes, R.O., An *in situ* interstitial water sampler for use in unconsolidated sediments, *Deep Sea Res.*, 20, 1125, 1973.

Bauer, J.E., Montgana, P.A., Spies, R.B., Prieto, M.C., and Hardin, D., Microbial biogeochemistry and heterotrophy in sediments of a hydrocarbon seep, *Limnol. Oceanogr.*, 33, 1493, 1988.

Bender, M., Martin, W., Hess, J., Sayles, F., Ball, L., and Lambert, C., A whole-core squeezer for interfacial pore water sampling, *Limnol. Oceanogr.*, 32, 1214, 1987.

Bischoff, J.L., Randall, E., and Luistro, A.O., Composition of interstitial waters of marine sediments: temperature of squeezing effect, *Science*, 167, 1245, 1970.

Bolliger, R., Brandl, H., Hohener, P., Hanselmann, K.W., and Bachofen, R., Squeeze-water analysis for the determination of microbial metabolites in lake sediments — comparison of methods, *Limnol. Oceanogr.*, 37, 448, 1992.

Bottomley, E.Z. and Bayly, I.L., A sediment pore water sampler used in root zone studies of the submerged macrophyte, *Myriophyllum spicatum*, *Limnol. Oceanogr.*, 29, 671, 1984.

Bray, J., Bricker, J.T., and Troup, O.P., Phosphate in interstitial waters of anoxic sediments, *Earth Planet. Sci. Lett.*, 18, 1362, 1973.

Brinkman, F.E., van Raaphorst, W., and Lyklema, L., *In situ* sampling of interstitial waters from lake sediments, *Hydrobiologia*, 92, 659, 1982.

Carignan, R., Interstitial water sampling by dialyses: methodological notes, *Limnol. Oceanogr.*, 29, 667, 1984.

Carignan, R., Rapin, F., and Tessier, A., Sediment pore water sampling for metal analysis: a comparison of techniques, *Geochim. Cosmochim. Acta*, 49, 2493, 1985.

Davey, E.W. and Soper, A.E., Apparatus for the *in situ* concentration of trace metals from sea water, *Limnol. Oceanogr.*, 20, 1019, 1975.

Davison W. and Zhang, H., *In situ* speciation measurements of trace components in natural waters using thin-film gels, *Nature*, 367, 546, 1994.

Davison, W., Grime, G.W., Morgan, J.A.W., and Clarke, K., Distribution of dissolved iron in sediment pore waters at submillimetre resolution, *Nature*, 352, 323, 1991.

Edmunds, W.M. and Bath, A.H., Centrifuge extraction and chemical analysis of interstitial waters, *Environ. Sci. Technol.*, 10, 467, 1976.

Elderfield, H., Caffrey, R.J., Leudtke, N., Bender, M., and Truesdale, V.W., Chemical diagenesis in Narragansett Bay sediments, *Amer. J. Sci.*, 281, 1021, 1981.

Emerson, S., Early diagenesis in anaerobic lake sediments; chemical equilibria in interstitial waters, *Geochim. Cosmochim. Acta.*, 40, 925, 1976.

Engler, R.M., Brannon, J.M., Rose, J., and Bigham, G., A practical selective extraction procedure for sediment characterization, in *Chemistry of Marine Sediments*, Yen, T.F., Ed., Ann Arbor Science Publisher, M.J., 1977, 163.

Fanning, K.A. and Pilson, M.E.Q., Interstitial silica and pH in marine sediments: some effects of sampling procedures, *Science*, 173 1228, 1971.

Froelich, P.N., Klinkhammer, G.P., Bender, M.L., Luedtke, N.A., Heath, G.R., Cullen D., Dauphin, P., Hammond, D., Hartmann, B., and Maynard, V., Early oxidation of organic matter in pelagic sediments of the eastern equatorial Atlantic: suboxic diagenesis, *Geochim. Cosmochim. Acta*, 43, 1075, 1979.

Goodman, K.S., An apparatus for sampling interstitial water throughout tidal cycles, *Hydrobiol. Bull.*, 13, 30, 1979.

Hartmann, M., An apparatus for recovery of interstitial water from recent sediments, *Deep Sea Res.*, 12, 225, 1965.

Hertkorn-Obst, U., Wendeler, H., Feuerstein, T., and Schmitz, W., A device for sampling interstitial water out of rivers and lake beds, *Environ. Technol. Lett.*, 3, 263, 1982.

Hesslein, R.H., An *in situ* sampler for close interval pore water studies, *Limnol. Oceanogr.*, 21, 912, 1976.

Höpner, T., Design and use of a diffusion sampler for interstitial water from fine grained sediments, *Environ. Technol. Lett.*, 2, 187, 1981.

Howes, B.L, Daecey, J.W.H., and Wakeham, S.G., Effects of sampling technique on measurements of pore water constituents in salt marsh sediments, *Limnol. Oceanogr.*, 30, 221, 1985.

Hursthouse, A.S., Iqbal, P.P., and Denman, R., Sampling interstitial waters from intertidal sediments: an inexpensive device to overcome an expensive problem?, *Analyst*, 118, 1461, 1993.

Jahnke, R.A., A simple, reliable and inexpensive pore water sampler, *Limnol. Oceanogr.*, 33, 483, 1988.

Kalil, E.K. and Goldhaber, M., A sediment squeezer for removal of pore waters without air contact, *J. Sed. Petrol.*, 43, 553, 1973.

Kepkay, P.E., Cooke, R.C., and Bowere, A.S., Molecular diffusion and the sedimentary environment: results from the *in situ* determination of whole sediment diffusion coefficients, *Geochim. Cosmochim. Acta.*, 45, 1401, 1981.

Kinnlburgh, D.G. and Miles D.L., Extraction and chemical analysis of interstitial water from soils and rocks, *Environ. Sci. Technol.*, 17, 362, 1983.

Lyons, W.B., Gaudette, H.E., and Smith, G.M., Pore water sampling in anoxic carbonate sediments: oxidation artifacts, *Nature*, 277, 48, 1979.

Mangelsdorf, P.C. and Wilson, T.R.S., Potassium enrichments in interstitial waters of recent marine sediments, *Science*, 165, 171, 1969.

Manheim, F.T., Interstitial waters of marine sediments, in *Chemical Oceanography*, Riley, J.P. and Chester, R., Eds., Academic Press, London, 1976, 115.

Matisoff, G., Lindsay, A.H., Matis, S., and Soster, F.M., Trace metal mineral equilibria in Lake Erie sediments, *J. Great Lakes Res.*, 6, 353, 1980.

Mayer, L.M., Chemical water sampling in lakes and sediments with dialysis bags, *Limnol. Oceanogr.*, 21, 909, 1976.

Montgomery, J.R., Price, M.T., Holt, J., and Zimmermann, C., A close interval sampler for collection of sediment pore waters for nutrient analyses, *Estuaries*, 4, 75, 1981.

Morrison, G.M.P., Bioavailable metal uptake rate determination in polluted waters by dialysis with receiving resins, *Environ. Technol. Lett.*, 8, 393, 1987.

Mueller, B. and McNee, J., *Design of "Aluminum Benthic Landers,"* University of British Columbia, Vancouver, 1994.

Murray, J.W. and Grundmanis, V., Oxygen consumption in pelagic marine sediments, *Science*, 209, 1527, 1980.

Presley, B.J., Brooks, R.R., and Kappel, H.M., A simpler squeezer for removal of interstitial water from ocean sediments, *J. Marine Res.*, 25, 355, 1967.

Reeburgh, W.S., An improved interstitial water sampler, *Limnol. Oceanogr.*, 12, 163, 1967.

Reeburgh, W.S. and Erickson, R.E., A "dipstick" sampler for rapid, continuous chemical profiles in sediments, *Limnol. Oceanogr.*, 27, 556, 1982.

Rey, J.R., Shaffer, J., Kain, T., Stahl, R., and Crossman, R., Sulfide variation in pore and surface waters of artificial salt-marsh ditches and a natural tidal creek, *Estuaries*, 15, 257, 1992.

Robbins, J.A. and Gustinis, J., A squeezer for efficient extraction of pore water from small volumes of anoxic sediment, *Limnol. Oceanogr.*, 21, 905, 1976.

Rosa, F. and Azcue, J.M., *Peeper Methodology — A Detailed Procedure from Field Experience*, Canadian Center for Inland Waters Report 93-33, 1993.

Rosa, F. and Davis, K., *Design of the "Quad-Clamp" Apparatus for Sediment Squeezing*, Canadian Center for Inland Waters, 1993.

Saager, P.M., Sweertz, J.P., and Ellermeijer, H.J., A simple pore water sampler for coarse, sandy sediments with low porosity, *Limnol. Oceanogr.*, 35, 747, 1990.

Sasseville, D.R., Takacs, A.P., and Norton, S.A., A large-volume interstitial water sediment squeezer for lake sediments, *Limnol. Oceanogr.*, 19, 1001, 1974.

Savile, H. and Pedrosa, M., *Optically Triggered Soft Sediment Placing Device ("Peeper Placer")*, Canadian Center for Inland Waters, ES-1095, 1980.

Sayles, F.L., The composition and diagenesis of interstitial solutions. I. Fluxes across the seawater-sediment interface in the Atlantic Ocean, *Geochim. Cosmochim. Acta.*, 43, 527, 1979.

Sayles, F.L., Mangelsdorf, P.C., Wilson, T.R.S., and Hume, D.N., A sampler for the *in situ* collection of marine sedimentary pore waters, *Deep Sea Res.*, 23, 259, 1976.

Sayles, F.L., Wilson, T.R.S., Hume, D.N., and Mangelsdorf, P.C., In situ sampler for marine sedimentary pore waters: evidence for potassium depletion and calcium enrichment, *Science,* 181, 154, 1973.

Scholl, D.W., Techniques for removing interstitial water from course coarse-grained sediments for chemical analysis, *Sedimentology*, 2, 156, 1963.

Siever, R., A squeezer for extracting interstitial water from modern sediments, *J. Sed. Petrol.*, 32, 329, 1962.

Simon, N.S., Kennedy, N.M., and Massoni, C.S., Evaluation and use of a diffusion-controlled sampler for determining chemical and dissolved oxygen gradients at the sediment-water interface, *Hydrobiologia*, 126, 135, 1985.

Troup, B.N., Bricker, O.P., and Bray, J.T., Oxidation effect on the analysis of iron in the interstitial water of recent anoxic sediments, *Nature*, 249, 237, 1974.

van Eck, G.T.M. and Smits, J.G.C., Calculation of nutrient fluxes across the sediment-water interface in shallow lakes, in *Sediment and Water Interaction*, Sly, P.G., Ed., Springer-Verlag, New York, 1986, 293.

van Raaphorst, W. and Brinkman, A.G., The calculation of transport coefficients of phosphate and calcium fluxes across the sediment-water interface, from experiments with undisturbed sediment cores, *Water Sci. & Technol.*, 17, 941, 1984.

Watson, P.G. and Frickers, T.E., A multilevel, *in situ* pore water sampler for use in intertidal sediments and laboratory microcosms, *Limnol. Oceanogr.*, 35, 1381, 1990.

Whiticar, M.J., Determination of interstitial gases and fluids in sediment collected with an *in situ* sampler, *Anal. Chem.*, 54, 1796, 1982.

Zimmermann, C.F., Price, M.T., and Montgomery, J.R., A comparison of ceramic and Teflon® *in situ* samplers for nutrient pore water determinations, *Estuarine Coastal Mar. Sci.*, 7, 93, 1978.

CHAPTER 5

Measurements and Handling of Samples in the Field

5.1 INTRODUCTION

The measurements and handling of sediment samples that should be carried out in the field are outlined in Figure 5-1. Only some of these operations will be required in many studies. However, when they are required, they should be performed immediately after retrieval of the sediment sample. In most studies relevant to dredging, measurements of Eh and pH, determination of cation exchange capacity and oxygen-free subsampling are not required. Therefore, preparing a proper description of the retrieved samples and mixing and subsampling them — described in Sections 5.3 and 5.6 of this chapter, respectively — will be sufficient in these studies.

Sediment from some areas can be contaminated by toxic compounds. However, this is usually not well known prior to sampling and analyzing the sediments. Therefore, we recommend wearing gloves and proper clothing to prevent skin contact during sediment handling. On the other hand, contamination of sediment samples by any protective clothing material must be avoided. Toxic fumes can be another problem when handling contaminated sediments. An open space during the field work and a fume hood during sediment handling in the laboratory are necessary to avoid inhaling the fumes. Goggles and a dust mask should be used to prevent irritation of eyes and inhalation of dust from dried sediments or splashed sediment particles during the mixing of wet sediments.

5.2 MEASUREMENT OF pH AND Eh

A quantitative interpretation of Eh measurements in natural aqueous systems is difficult because of problems associated with the measurement technique, the performance of the inert metal electrode, and the thermodynamic behavior of the environment (Whitfield, 1969). Critiques regarding the use of

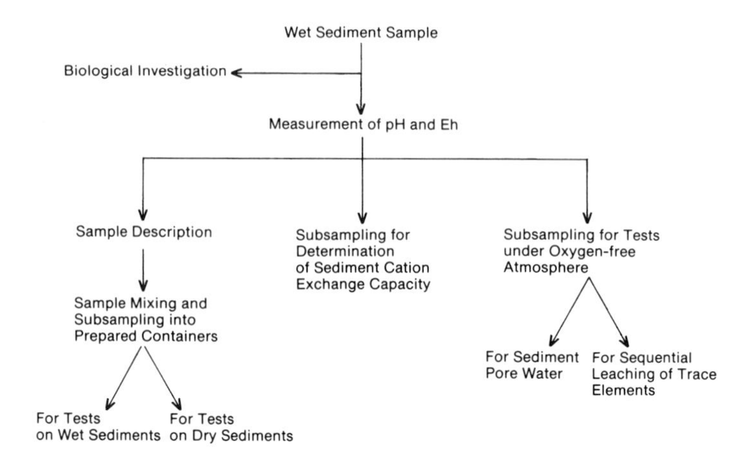

Figure 5-1 Sample handling and measurements in the field.

Eh for determining the redox level of marine sediments have pointed out that the absence of uniform measurement procedure (choice and preparation of electrodes, their retention time in the sediment, etc.) leads to different results (Rozanov, 1982). There are other additional problems associated with measuring pH and particularly Eh in sediments.

In the past, the potential of the platinum electrode has been used to describe oxidation-reduction conditions in bottom sediments. However, there are several limitations on the use of the platinum electrode in measuring Eh in the sediments. Whitfield (1974) pointed out several main physico-chemical objections to the use of Eh as an operational parameter. Some problems identified by Whitfield and others observed during our attempts to measure Eh in bottom sediments are as follows:

- Difficulties associated with the disturbance of the sediment sample, resulting in release or adsorption of gases, particularly O_2 and H_2S, and reactions at the liquid junction of the reference electrode, such as precipitation of metal sulfides. The disturbance of the sediment sample can occur during sampling; however, it mainly originates during the insertion of the electrode into the sediment.
- Instability and poor reproducibility of the measurements originating from low exchange current densities at the platinum surface and the predominance of mixed potentials. Measurements obtained by the platinum electrode depend on the nature and rates of the reactions at the electrode surface. Therefore, a quantitative interpretation of the measured values is very difficult.

- Response of the platinum electrode to changes in the environment will depend greatly on the properties of the platinum surface and the presence of adherent surface coatings. For example, upon inserting the electrode into fine-grained sediments containing a small amount of O_2 and reducing agents, a slow formation of platinum sulfide on the surface of the electrode may generate more negative potentials. On the other hand, in coarser sediments from oxygenated waters, such as mixtures of sand and silt in a nearshore area of a lake, the platinum may act as an oxide electrode and respond to pH rather than to the partial pressure of oxygen.

The above-listed problems indicate that the potential measured in the sediment samples is not a reduction-oxidation potential but rather a response of the electrode used in the measurements. Studies on the meaning of and problems with measuring and interpretation of sediment pH and Eh (Mortimer, 1941, 1942; Berner, 1963; Morris and Stumm, 1967; Whitfield, 1969; Tinsley, 1979; Bates, 1981) should be consulted prior to measuring these parameters in sediment samples. Further, we recommend consideration of the excellent reviews by Berner (1981) and Thorstenson (1984) on the implications of several factors of electron activity for the study and interpretation of redox processes in sedimentary environments and natural waters. For those who decide to measure the Eh after considering the above limitations, some examples from studies involving Eh measurements and recommendations on procedure for field measurements are given below.

For the Eh measurement, Whitfield (1969) used two platinum electrodes immersed side by side in the sediment sample to ensure that reproducible potentials were being measured. The values of Eh measured by the two electrodes differed by 10 to 30 mV due to the surface properties of the platinum and poisoning of the platinum surface by an irreversible attack at different rates to the two electrode surfaces. To measure simultaneously Eh, pH, and sulfide activity (pS^{-2}), five electrodes were combined into a compound electrode, enabling all working electrodes to be introduced in a single probe into the sediment. When this probe was used in conjunction with a remote junction reference electrode, stable and reproducible results were obtained, even on highly reduced samples. The Eh of each sample was monitored until a steady value (drift less than 1 mV/min) was observed. This value was recorded and the Eh measured by the second of the five electrodes was registered. The sulfide activity and pH were measured, and the Eh registered by the first electrode was checked again. When the results differed by more than 30 mV, a further 5 minutes were allowed for equilibration. The temperature was recorded before the sample was discarded. The whole cycle of readings was completed within approximately 10 minutes.

Whitfield found that cleaning the platinum electrodes by rubbing with a fine abrasive cloth was more efficient than rinsing with various chemicals and distilled water. The potential of the Eh electrode/calomel cell was measured occasionally in a Zobell solution (Zobell, 1946) — 0.003 M potassium ferricyanide and 0.003 M potassium ferrocyanide in 0.1 M potassium chloride — to check the performance of the liquid junction. The potential of the cell was adjusted to the standard hydrogen scale by adding 250 mV to the measured value to enable comparison with other data.

Bagander and Niemisto (1976) measured the Eh in marine sediment cores with a specially designed electrode inserting an attachment adaptable to a subsample slide used in the sectioning of the cores. This attachment allowed electrode measurements with minimal air contamination and disturbance of the subsamples. Two different types of electrodes were used: one type comprised a platinum wire fused into the end of a glass tube and an Ag/AgCl reference electrode; the other was an Orion combination electrode.

A portable Metrohm pH and Eh meter and combination glass and platinum electrodes were used for measurements of pH and Eh in surficial sediments and sediment cores collected from the Laurentian Great Lakes (Kemp and Lewis, 1968; Kemp et al., 1971). Surface sediment samples were collected by a Shipek grab sampler. Upon retrieval, the sampling bucket with the collected sediments was placed on a special stand to keep the sediment surface horizontal. The Eh electrodes were calibrated using a Zobell solution (Zobell, 1946) and marked vertically at 0.5-cm intervals. They were then pushed into the sediment to the desired depth. The pH electrodes were similarly marked and were calibrated using two buffer solutions of a known pH value (4.0 and 7.0). For measuring pH and Eh in sediment cores, the electrodes were inserted into the sediment after the core was placed on an extruder, uncapped, and water from the top of the core siphoned off. For measuring the pH and Eh of the top 1 cm sediment layer, the electrodes were pushed 0.5 cm into the sediment. The pH reading was taken about 1 minute after inserting the electrode into the sediment. However, it took approximately 10 minutes to stabilize the Eh value on the meter. After measuring the first subsample, the electrodes were removed from the sediment, cleaned with distilled water, and dried with a soft paper tissue. The sediment layer for which the measurements were made was subsampled into a prepared container and the electrodes were inserted into the next layer of the sediment core. After every five measurements, each electrode was recalibrated.

During hundreds of pH measurements of estuarine and normal marine sediments, Berner (1981) never encountered pH values outside the range of 6 to 8. In more than 90% of the measurements he or his colleagues carried out, the pH was within the range of 6.5 to 7.5. Similarly, the pH we measured in lacustrine sediments such as the Laurentian Great Lakes and other lakes in Canada ranged mainly from 6.5 to 7.3. This information indicates that in most cases pH is not a suitable parameter for characterization of bottom sediments.

A general description of the procedure for measuring pH and Eh in bottom sediments is given below. However, the instructions given by the manufacturer of the electrodes used in the measurements should be considered in the preparation of the electrodes prior to measuring pH and Eh in sediment samples.

5.2.1 General Procedure for Measurements of pH and Eh in Bottom Sediment

The pH and Eh in sediment samples should always be measured immediately upon retrieval of the undisturbed samples to avoid the effect of changes in sediment chemistry on measured values. Therefore, in most cases the measurements must be carried out in the field.

Equipment and Solutions Used in the Measurements

- A portable, battery-operated pH/Eh meter, batteries, and a power cord for recharging the meter.
- Combination glass and platinum electrodes or other electrodes suitable for the measurements.
- Plastic test-tube-shaped containers with rubber sleeves or glass containers with ground joint for storing the electrodes in solutions during transport and in the field. These containers are sometimes supplied by the electrode manufacturer. They also can be made from a test tube of a suitable size.
- Commercially available or laboratory-prepared pH buffer solutions (pH 4 and 7) in plastic bottles with lids.
- Freshly prepared solution for calibration of Eh electrode, such as a Zobell solution (0.003 M potassium ferricyanide and 0.003 M potassium ferrocyanide in 0.1 M potassium chloride) in a plastic bottle with a tight lid. The bottle with the solution *must* be labelled and handled according to the safety regulations for cyanides.
- Freshly prepared solution of saturated potassium chloride for storage of the electrodes and maintaining the solution inside the combination electrodes. The solution should be stored in a plastic bottle with a lid.
- Other solutions necessary for the proper functioning of the electrodes during the field measurements, as outlined by the manufacturer of the electrodes.
- Distilled water and wash bottle for storing and rinsing the electrodes between and after the measurements. The quantity of the distilled water will depend on the number of measurements.
- Small beakers for holding the buffer and calibration solutions and larger beakers or other containers for rinsing the electrodes after the measurements. Plastic beakers or containers are more convenient to handle in the field than glass beakers or containers.
- Support stands, rods, connectors, and clamps to secure the electrodes in buffer and calibration solutions, in the sediment samples during the measurements, and during cleaning.
- Large plastic containers for storage and transport of used buffers and Eh-calibration solutions. Used solutions need to be properly discharged according to the regulations for discharge of cyanide-containing liquids.
- Note book and pens, soft paper tissue.
- Combination electrodes should be stored according to the manual provided by the manufacturer. In most cases, combination glass electrodes should be stored in saturated potassium chloride solution when not in use to prevent the glass membrane from drying out.

The above-described equipment is shown in Figure 5-2.

Figure 5-2 Equipment used in measuring sediment pH and Eh.

Preparation of the Equipment Before the Field Trip

- Check the batteries of the portable pH/Eh meter and recharge them, if necessary.
- Prepare the solutions for calibration of the Eh and pH electrodes in the field.
- Check and test the pH and Eh electrodes. Check the electrodes' connections to the meter, remove them from the storage solution, wash them with distilled water, and calibrate them with the buffer and other solutions.
- Mark the electrodes vertically at desired intervals for insertion into the sediment samples.
- Store the electrodes, according to the manufacturers instructions or in saturated potassium chloride solution, for transport and use in the field.
- Pack all equipment and solutions for transport to the field.
- Take spare electrodes to the field if available.

Measurements in the Field

- We recommend allocating a space in the field where all pH and Eh measurements will be carried out. Within this space, all equipment should be assembled, checked for proper functioning, and prepared for measurement of the first sample. Electrodes have to be connected, checked, calibrated, washed, and prepared for measurements of pH and Eh prior to the retrieval of the sediment sample.
- Grab sampler and sediment corers with recovered sediment need to be placed in such way that they will remain steady without disturbing the sediment samples during the measurements.
- Electrodes must be carefully inserted into the undisturbed sediment samples to avoid any air contamination, particularly around the Eh electrode. Care must be taken not to generate open space between the electrode and the sediment. Proper insertion of the electrode without disturbing the sediment is the most important step in measuring the Eh.

- The electrodes are inserted into the sediment to the depth marked. The pH/Eh meter is switched to the pH scale and the value recorded within 1 minute after inserting the electrode into the sample. The meter is then switched to the mV scale for recording the Eh value. The potential usually drifts considerably over the first 10 to 15 minutes, and after this period becomes stabilized. After stabilization, the value on the mV scale should be recorded. In measuring Eh of sediments from waters with low ionic strength, such as most freshwater bodies, it is recommended to "acclimatize" the electrodes in the water prior to measurement, particulary the electrodes that were stored in saturated potassium chloride solution. This will reduce drifting of the potential after inserting the electrode into the sediment. In our measurements we have used Zobell's solution for the calibration of the platinum electrode. The solution has an Eh value of +430 mV at 25°C. The Eh value of the Zobell solution recorded by the platinum electrode was usually around +250 mV. Therefore, the difference (i.e., +180 mV) was always added to the values measured in sediment samples by the electrode to standardize the measurements. However, we measured the Eh mainly to obtain general information on the oxic or anoxic conditions in the sediments with no further attempts to interpret the measured values.
- Both electrodes are then removed from the sample, washed with distilled water to remove all adhering sediment particles, and dried gently with a soft paper tissue.
- The electrodes should be calibrated after each fifth measurement. However, in measuring pH and Eh in a sediment core the electrodes may need a less frequent calibration.

5.3 SAMPLE DESCRIPTION

All retrieved samples should be described after measuring pH and Eh. It will likely be necessary to split the sample with a spatula or knife to expose the inner sediment in the sampling bucket of a grab sampler. Such a disturbed sample is not suitable for measuring pH and Eh. Observations on sediments in a grab sampler and sediment cores should be recorded regarding texture, color, odor, presence of biota, and foreign matter, etc., as described in Section 6.4.6.

5.4 SUBSAMPLING FOR DETERMINATION OF CATION EXCHANGE CAPACITY

Cation exchange capacity (CEC) should be determined on wet, untreated bulk sediments immediately after sample collection. Therefore, sediment subsamples collected for the determination of CEC should be kept at 4°C and analyzed as soon as possible. Adams et al. (1980) studied the effects of freeze-drying, size fractionation, organic matter removal with 30% H_2O_2, and colloidal iron removal with citrate-dithionate for changes in sediment CEC, and found from 71% increase to 40% reduction of CEC by these different treatments.

5.5 SUBSAMPLING UNDER OXYGEN-FREE ATMOSPHERE

Chemical species of trace elements and their association with different sediment components have been investigated in environmental studies and in geochemical exploration (Gupta and Chen, 1975; Tessier et al., 1979, 1982). The concentrations of different chemical forms of trace elements associated with sediment components can be used to predict their bioavailability in sediments (Jenne and Luoma,1977). Analyses of sediment pore water are used for developing thermodynamic models to determine the partitioning of sediment-associated major and trace elements in sediment/water systems (Carignan et al., 1985).

Results of these investigations can be influenced by the degree to which the integrity of the sediment sample is preserved between the time of collection and chemical analyses. Rapin et al. (1986) tested the effects of sample handling and storage on the results of a sequential extraction procedure for determining the partitioning of trace metals in freshwater sediments. Their study showed that drying, freeze-drying, or oven-drying should be avoided, and that freezing or short-term wet storage (at 1–2°C) are acceptable preservation techniques. Of the nine metals tested (cadmium, cobalt, chromium, copper, nickel, lead, zinc, iron, and manganese), chemical species of copper, iron, and zinc in sediments were particularly sensitive to sample handling.

The maintenance of oxygen-free conditions during subsampling and extraction of anoxic sediments is of critical importance. Consequently, the preservation of sediment constituents from the effects of the air and biological activity is a fundamental requirement. Except for the surface 1- to 3-cm layer, bottom sediments are usually anoxic and become rapidly oxidized by air. Suspended sediments collected from the water column are usually oxidized.

Therefore, the handling and preparation of bottom sediments collected for determining chemical forms of trace elements and collecting sediment pore water must be carried out in an oxygen-free atmosphere. Bulk sediment samples collected by a grab sampler should be transferred into storage containers immediately upon retrieval. Head space should be avoided by filling containers to the top and sealing air-tight. Sediment cores must be extruded in an oxygen-free atmosphere using a glove box (Figure 5-3) or plastic bag filled with an inert gas such as nitrogen. Glove boxes are available from suppliers of laboratory equipment. A simple piston extruder similar to that shown in Figure 5-7 can be used for subsampling sediment cores in a glove box. The plastic liner

Figure 5-3 Glove box for handling sediment samples in an oxygen-free atmosphere.

with the recovered sediment core must be fitted through an opening cut in the bottom of the glove box. Before commencing sample handling and core subsampling, the air in the glove box must be replaced by a constant, controlled volume of inert gas supplied from a cylinder (see Figure 5-3). We recommend placing an oxygen meter in the glove box to ensure a completely oxygen-free atmosphere.

The sediment core is subsampled and individual sections are transferred into desired containers and tightly closed in the glove box. When pH and Eh measurements are required, they should be carried out in the glove box prior to or during subsampling by inserting the electrodes into individual sections before sectioning the core and placing the subsamples into containers. For collection of sediment pore water, sediment is often transferred into centrifuge tubes in the glove box. After centrifugation, the tubes are returned to the glove box to remove the pore water and prepare the samples for analysis. If a sequential extraction procedure is undertaken to determine chemical forms of trace elements, all sediment handling and extraction steps should be carried out in the glove box. This includes deaeration, withdrawal and/or addition of extraction solutions, and sealing the extracts. For example, Filipek and Owen (1979) carried out the sequential extraction procedure on different particle size fractions on sediments collected in Lake Michigan. The grain size separation was achieved by wet-sieving well-mixed subsamples of sediments (20 g) in a glove box under a nitrogen atmosphere to avoid changes in the chemical forms of the metals in the sediments.

Generally, sample handling and preparation for determining chemical species of metals and trace elements in sediments using a sequential extraction procedure should always be carried out under conditions described in the pertinent scientific literature or established protocols.

5.6 SAMPLE MIXING AND SUBSAMPLING INTO PREPARED CONTAINERS

5.6.1 Subsampling of Grab Samplers

Grab samplers can collect up to a 15-cm-deep layer of bottom sediments. The collected sample usually contains about 2 to 5 cm of soft, fine-grained, oxidized surface sediments with a high water content and underlying, reduced, more consolidated sediment. The sample can also contain a heterogeneous mixture of sand and silty clay. Therefore, in a case where the whole sediment sample collected by a grab sampler needs to be divided into subsamples for different analyses, the sediment must be manually homogenized in the field by mixing in sufficiently large containers. An intensive manual mixing of wet sediment is usually sufficient to homogenize the sample prior to subsampling into containers for analysis (Figure 5-4). Sediments can be transferred from the grab sampler to a precleaned glass or stainless steel bowl and thoroughly homogenized by stirring with a glass or stainless steel spatula or scoop until textural and color homogeneity are achieved (Chapman, 1988). In this procedure, the entire content of a grab sampler has to be carefully deposited into a clean, inert container of a similar size and shape as the sampler upon retrieval of the sample. We recommend placing the grab sampler into the container and slowly opening the jaws of the sampler to allow deposition of the sediment sample into the container. The sediment in the container is then homogenized and divided into individual precleaned, labelled containers. However, it has to be recognized that the sediments become oxidized rapidly by compositing and homogenization. Therefore, samples obtained by this method are not suitable for studies requiring undisturbed sediment, such as sediment pore water studies, determination of chemical forms of metals, etc. Landrum et al. (1989) sieved collected sediments through a 1-mm sieve until they appeared to be homogeneous prior to studying the availability of contaminants. Breteler et al. (1989) homogenized sediments collected for bioassays by combining five grab samples in a Teflon®-coated stainless steel pan and stirring thoroughly with a Teflon®-coated stainless steel spoon. Homogenized sediment was then added to half-gallon polypropylene bottles for chemical analyses and bioassays.

In some studies, the topmost sediment, such as the surficial 2 or 3 cm sections, need to be subsampled from a grab sampler. Therefore, good access to the surface of the sediment collected by the grab sample without any loss of the top fine-grained sediments is absolutely necessary. Selecting the correct type of grab sampler to fulfil this requirement is very important. Suitable grab samplers for collection of the topmost sediments are those with removable buckets, such as Shipek grab, or opening flaps or screens on their top (Figure 5-5). Water on the top of the retrieved sediment in the grab sampler must be carefully siphoned off without disturbing the sediment surface. This can be achieved by siphoning with plastic tubing in a manner similar to siphoning

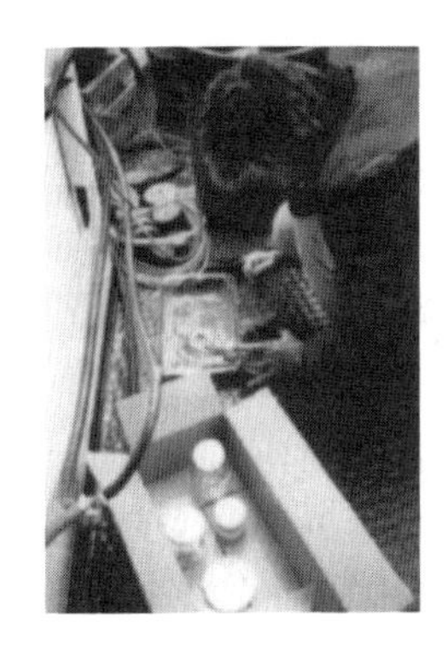

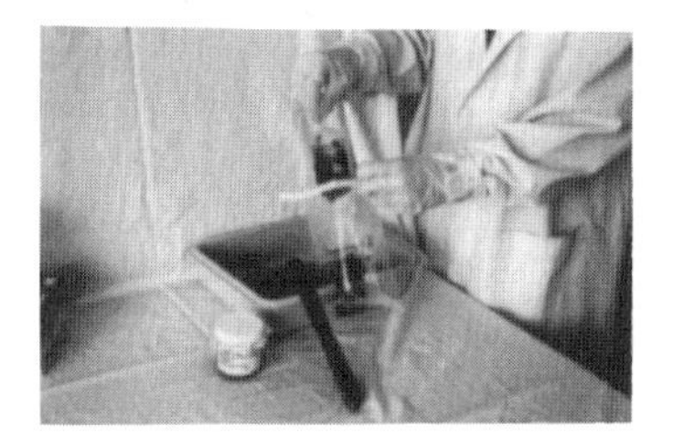

Figure 5-4 Homogenization of sediments by mixing and dividing into containers for analysis on the ship deck and in the laboratory.

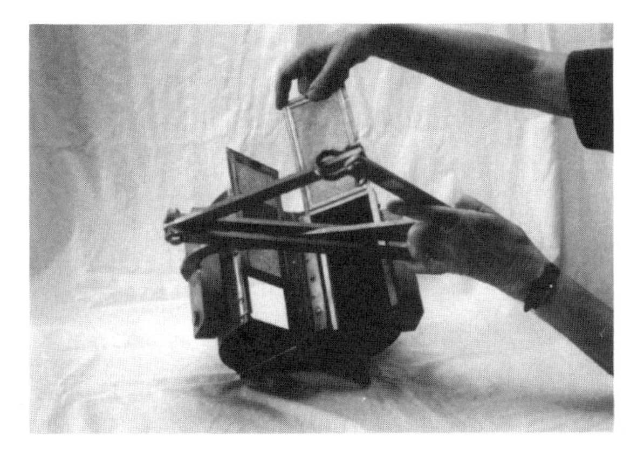

Figure 5-5 Opening the top of a grab sampler.

water from a sediment core, as shown in Figure 5-8. A clean Teflon® spoon or scoop is suitable for subsampling the topmost sediment from the grab sampler. It is also possible to use a small core tube that is pushed gently into the sediment in the grab sampler to the desired depth to be subsampled. A Teflon® plate is then inserted below the tube to retain the sediment inside the tube. The core tube with the sediment is then pulled from the rest of the sediment in the grab sampler, the sediment adhering to the outside of the tube and the plate is washed off, and the sediment inside the tube is collected into a sample container. When larger quantities of surface sediments need to be collected by subsampling, the subsamples collected from several grab samplers need to be composited. The compositing can be carried out in a large, prewashed Teflon®, glass, or porcelain dish using a spoon made from the same material. Individual subsamples obtained by repeated sediment sampling from a grab sampler are collected in the dish and homogenized by hand-mixing using a spoon. The homogenized sample is then divided into prewashed, labelled containers (Figure 5-4).

If all analyses of sediments are to be performed on dry sediment, the bulk wet sediment sample should not be subsampled in the field, but should be dried as a bulk sample and further processed in a dry state (see Section 6.4). Containers and implements used for homogenization and subsampling should be chosen with the same considerations as for the sample containers described in Section 6.2.

5.6.2 SUBSAMPLING OF SEDIMENT CORES

Sediment cores must be stoppered immediately after retrieval to prevent accidental loss of samples (Figure 5-6). Recovered sediment cores must be processed with respect to the type and number of analyses that need to be carried out on each core and as outlined in the sediment sampling plan. The cores should be subsampled as soon as possible after their retrieval. When using a clear plastic liner, the appearance of the sediment core should be recorded prior to subsampling, along with other features such as the length of the core; sediment color, texture, and structure; occurrence of fauna; etc.

Long cores, such as those collected by piston coring, can be cut into sections of suitable length for storage, sectioned longitudinally, described, labelled, wrapped to preserve original consistency, and transported for storage in a refrigerated room. Sediment cores collected for stratigraphical or geotechnical studies can be stored at 4°C in a humidity-controlled room without any large changes in sediment properties for several months.

Sediment cores collected for chemical analyses, particularly for determination of contaminants, should be extruded from the core liners and subsampled as soon as possible. Cores collected with gravity corers are usually up to 2 m long and have to be kept upright to prevent mixing of the uppermost part of the sediment core, which usually consists of very fine, soft, and unconsolidated material. Prior to any transport of these cores, the entire space over the sediment in the core liner needs to be filled with lake or sea water, and both ends of the core liner have to be completely sealed to prevent mixing of the sediment inside the liner. A refrigerated storage space should be used in case the cores have to be stored, even for a short time. The cores should be stored in an upright position and secured from falling over. Sediments with a high content of organic matter often contain large amounts of gasses which, upon the recovery, can disturb a section of the core or the entire core.

Sediment cores collected for studies of environmental pollution or sediment dating are usually subsampled into 1-cm sections. By subsampling the sediment into larger sections, for example 3 to 10 cm thick, the information on the vertical distribution of contaminants can be lost, particularly in sediments collected from an area with a low sedimentation rate. For example, it would be impossible to assess recent changes in contaminant loadings from 5-cm core sections at an area with a 2-mm annual sedimentation rate.

Often, budgetary limitations will determine the number of analyses and thus the decision on the number of subsampled core sections. To limit a large number of analyses, we recommend initially analyzing every third section of a core subsampled into 1-cm sections, while storing the remainder for additional analyses when required.

There are various methods for subsampling sediment cores. Håkanson and Jansson (1983) described an electroosmotic knife and/or guillotine, dry-ice freezing method, and a Plexiglass slide for subsampling sediment cores. The cores can be extruded by a piston-type extruder (Kemp et al., 1971). Extruding and sectioning cores using a simple type of extruder involves the few steps outlined in Figures 5-7 to 5-11. The capped core liner containing the sediment

Figure 5-6 Stoppering a sediment core collected by a Benthos gravity corer.

Figure 5-7 Simple piston extruder for subsampling sediment cores with a metal rod for securing longer cores from falling over.

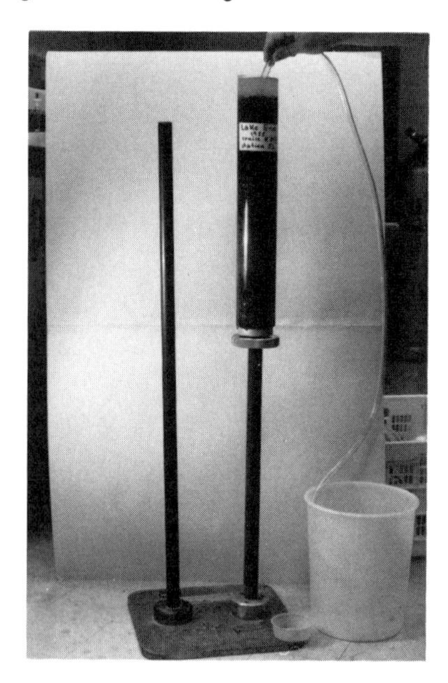

Figure 5-8 Siphoning off the water from a sediment core placed on an extruder piston before commencing subsampling.

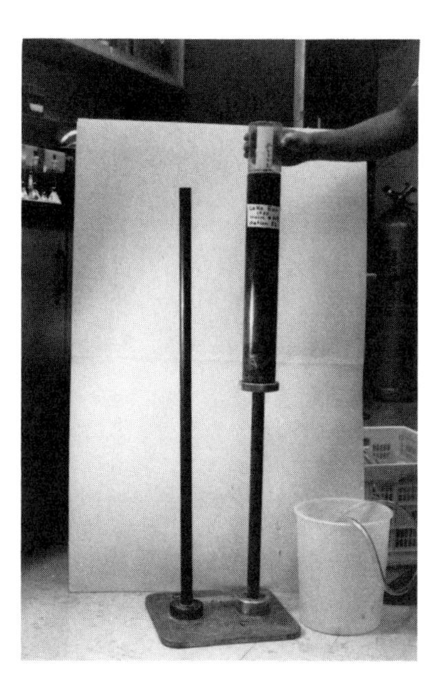

Figure 5-9 Placing a piece of core liner with a scale on the top of the core.

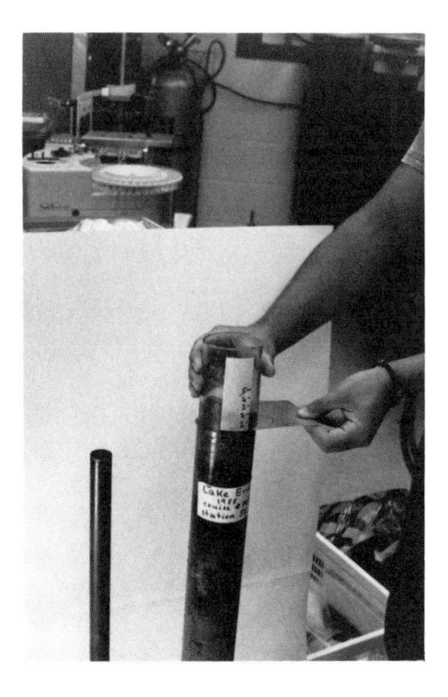

Figure 5-10 Sediment extruded into the piece of core liner with the scale cut by a metal cutter.

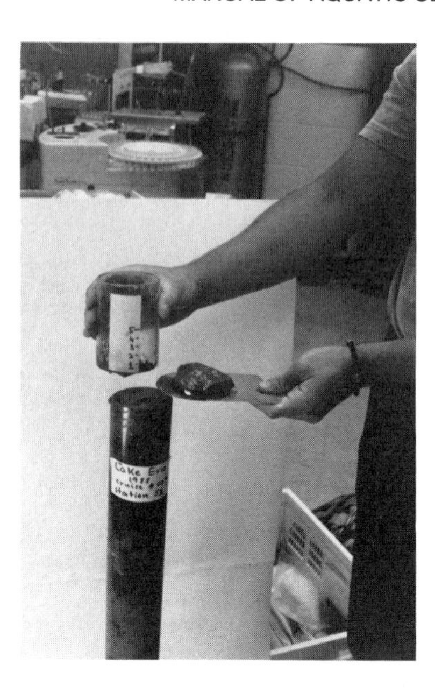

Figure 5-11 Sediment subsample on a metal cutter.

and overlying water is uncapped at the lower end and placed vertically on top of the piston. The top cap is removed and the water is siphoned off to avoid disturbance of the sediment-water interface. The core liner is then pushed slowly down until the surface of the sediment is at the upper end of the liner. Sediment sections are collected by pushing the liner down and cutting the exposed sediment into sections of the desired thickness. From each sediment section, a 1- to 2-mm outer layer of sediment that has been in contact with the plastic or metal liner is discarded to avoid contamination. A simple device to achieve this outside-layer removal has been constructed by the National Water Research Institute, Environment Canada, Burlington, Ontario. The device consists of a frying pan with a hole cut in the center to fit the plastic liner. A metal ring with a sharp bottom edge is mounted over the hole in the pan to cut only the center part of the sediment core (Figure 5-12). The use of the "frying pan" in the subsampling of sediment cores is shown in Figures 5-13 to 5-15. Individual core sections are collected into precleaned, labelled containers. Because the uppermost sediment layer often consists of very fine sediments with a high water content, it may be necessary to subsample using a pipette or a large syringe with a wide-open end to prevent blocking by sediment particles. The extruder shown in Figure 5-7 is suitable for subsampling approximately 2-m-long, fine-grained sediment cores collected with a gravity corer. Alternatively, cores of more consolidated material can be mounted onto a horizontal U-shaped rail and the liner cut using a saw mounted on a depth-controlling jig. The final cut can then be made with a sharp knife to avoid contamination of the sediment by liner material, and the core itself can be sliced with Teflon® or

Figure 5-12 Equipment ("frying pan") for subsampling sediment cores and eliminating the sediment that has been in contact with the core liner.

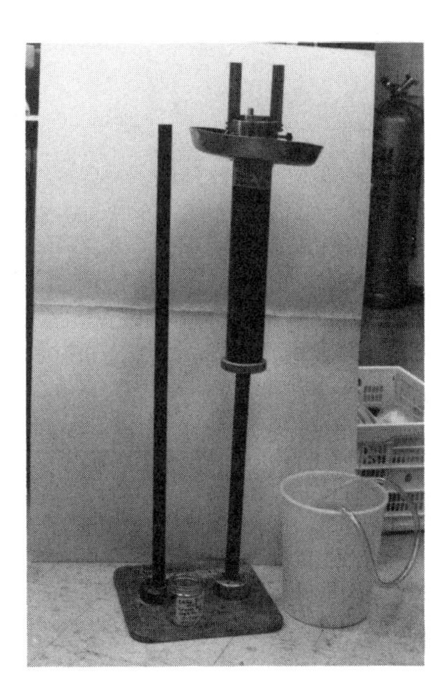

Figure 5-13 "Frying pan" placed on the top of the sediment core before subsampling.

nylon string. The core then becomes two D-shaped halves that can be easily inspected and subsampled.

The extruder, shown in Figure 5-16, was designed and constructed for subsampling sediment cores in the field at the National Water Research Institute, Environment Canada, Burlington, Ontario. It consists of approximately a

Figure 5-14 Cutting a subsample from the sediment core using the "frying pan."

Figure 5-15 Collecting a sediment subsample in a prewashed and labelled glass jar.

50 cm × 50 cm × 0.5 cm aluminum plate with a reinforced aluminum frame and two stainless-steel bases mounted on the surface of the plate. The inside diameter of one base corresponds to the outside diameter of the plastic liner commonly used with gravity corers, such as the Benthos corer, and the inside diameter of the second base corresponds to the outside diameter of the plastic liner used with the lightweight corer described by Williams and Pashley (1979). Therefore, this extruder can be used for subsampling sediment cores collected in these two different sizes of core liners. The core liner is clamped in place with two special holders cut to match the outside diameter of the core tube being used. This holds the core upright and secured in the base. A hose is connected to each base to fill the inside space of the base with water. The other end of the hose is connected to a water tap or, at areas with no water supply, to the metal pressurized stainless-steel bottle filled to three-quarters of

its volume with water under pressure. A stopper, made from two rubber stoppers and joined vertically by a screw, is inserted into the bottom end of the core liner upon retrieval of a sediment core. This rubber stopper fits completely into the plastic liner to prevent the sediment from sliding out. The core is then placed into the base on the extruder plate; the top cap is removed and the sediment pushed upward by water from the water tap or the pressurized bottle. The water pressure is controlled by a fine needle valve inserted in the supply hose. The sediment extruding from the top of the core liner can be subsampled into sections of desired thickness (minimum of about 0.5-cm-thick slices).

Freezing sediment cores selected for subsampling is not suitable. Sediment freezing changes the sediment volume depending on the water content, and 1 cm of a frozen (or frozen and thawed) sediment section does not equal 1 cm depth of fresh sediment.

Recently, a segmented gravity corer was described by Aanderaa Instruments, Victoria, British Columbia, Canada. The core tube of the sampler consists of a series of rings placed on top of one another (Figure 5-17). Subsampling is carried out by rotating the rings around its other axis so that it cuts sediments layer of similar thickness (Figure 5-18). The method is fast and accurate without disturbing the sediment in the core tube. Testing of the corer by the manufacturer indicated its suitability in sampling of fine-grained sediments. The design of the segmented core tube enabled easy subsampling of the core into 1-cm sections and their subsequent transfer into sample containers by one person in the field.

Figure 5-16 Core extruder using pressurized water.

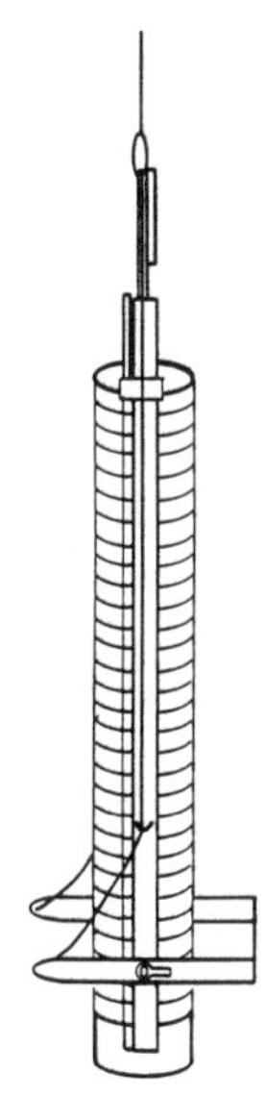

Figure 5-17 Segmented core tube of a gravity corer (courtesy of Aanderaa Instruments, Victoria, B.C., Canada).

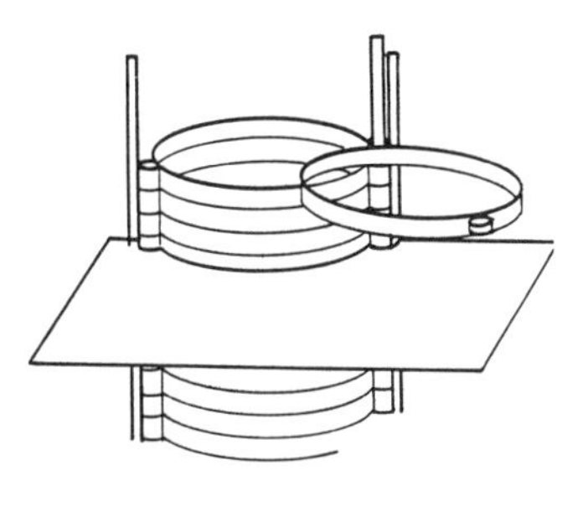

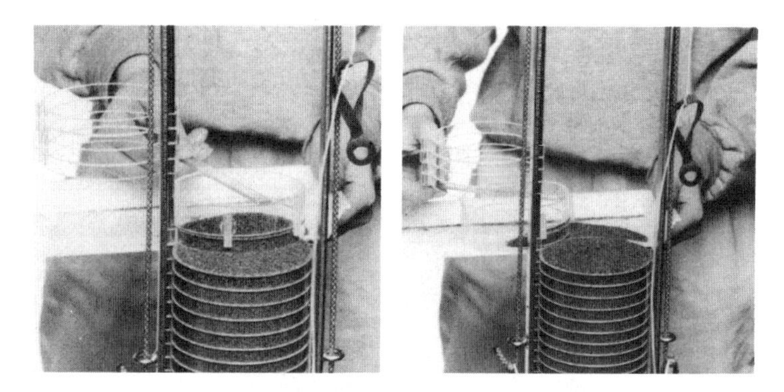

Figure 5-18 Sketch of the segmented gravity corer and subsambling of a sediment core collected by the segmented gravity corer (courtesy of Aanderaa Instruments, Victoria, B.C., Canada).

5.7 SAMPLING HAZARDOUS SEDIMENTS AND SAFETY REQUIREMENTS

The real challenge to society today is to manage and dispose of hazardous wastes, although they represent less than 5% of the total municipal waste. Hazardous waste is defined as "solid waste that may cause increased mortality or serious illness or may cause substantial hazard to human health or to the environment when improperly managed" (Shaheen, 1992). Hazardous waste sites pose a multitude of health and safety risks, any one of which could result in serious injury or death (Martin et al., 1992). These risks are due to the physical and chemical nature of the site as well as the work being performed. Hazardous sites contain chemicals that may be harmful to humans, including those who collect samples at these sites. Table 5-1 summarizes typical hazards and their causes encountered when sampling materials from hazardous waste sites or spills.

Planning is the first step in avoiding danger when sampling sediments from hazardous waste sites. Safety must always be considered in the development of any sampling plan, and only well-trained personnel should carry out this type of sampling. Correct planning and execution of safety protocols help protect sampling personnel from accidents and unnecessary exposure to hazardous chemicals. The study leader should obtain as much information as possible about the site characterization before personnel enter the site. At the site of a spill accident or when the hazards are largely unknown, we recommend first making visual observations and collecting samples near the site. Among the information that should be obtained before going to the site are:

- exact location of site and development of a preliminary site map;
- ownership and management of the site;
- necessary permission to enter the site;
- activities on the site;
- accessibility and easy exits from the site;
- a list of hazardous substances disposed at the site and their chemical and physical properties;
- geological, hydrological, and meteorological conditions at the site; and
- previous surveying, sampling, and monitoring data.

The off-site information should be used to develop a safety plan for sampling, addressing the work to be accomplished and duration, and prescribing the procedures to protect the health and safety of the sampling team. The composition of the sampling team depends on the site characteristics, but should usually consist of four persons, two who enter the site and sample and two outside support people outfitted with personnel protective equipment and ready to enter the site in case of emergency (Martin et al., 1992). Accurate, current, and readily accessible information about the site conditions, its man-

Table 5-1 Typical Hazards Encountered when Sampling Sediments from Hazardous Waste Sites or Spills

Hazard type	Exposure route or causes	Prevention
Chemical exposure	Inhalation, eye/skin contact, ingestion, puncture	Use remote sampling devices when possible
Ionization radiation	Sediments containing radioactive materials Exposure, radiation burns, ingestion	Perform radiation survey early in investigation Wear protective clothing and dust masks
Physical safety hazard	Steep grades Slippery surfaces Uneven terrain Sharp objects	Perform visual inspection and monitoring Wear better-fitting clothing Wear hard hats, boots with good gripping soles
Biological (etiological) hazards	Sediments containing wastes from hospitals and research facilities Poisonous plants, insects, animals, and indigenous pathogens	Wear gloves, respirators, protective clothing Decontaminate with disinfectant Immunize if agent is known
Heat stress or cold exposure	Work done in clothing designed to protect against chemicals but not against weather conditions	Take frequent rest breaks Monitor body temperature Drink fluids Wear appropriate clothing for weather conditions
Fire and explosion	Unstable chemicals Incompatible reactions Vapor buildup	Use nonsparking tools Use fire proximity suits or blast suits

agement, and past and present activities carried out at the site are essential for assessing hazards, reviewing plans, and making decisions regarding the sampling. Documentation may become crucial in the event of any litigation (see Section 7.5).

An additional problem for groups involved in sporadic sampling of hazardous sediments is the contamination of the sampling equipment. Martin et al. (1992) recommended the following alternatives to prevent or minimize contamination:

- Use remote sampling, handling, and container-opening techniques (e.g., drum grapplers, pneumatic impact wrenches).
- Minimize contact with hazardous sediments (e.g., avoid walking on or touching potentially hazardous areas).
- Protect sampling equipment by bagging. Make openings in the bags for sample ports and sensors that must contact the sediments or soils.
- Wear disposable outer garments and use disposable equipment where appropriate.

- Cover sampling equipment and tools with a coating that can be removed during decontamination.
- Prepare a contingency plan for all situations (accessibility to medical help, etc.).

REFERENCES

Adams, D.D., Darby, D.A., and Young, R.J., Selected analytical techniques for characterizing the metal chemistry and geology of fine-grained sediments and interstitial water, in *Contaminants and Sediments*, Vol. 2, Baker, R.A., Ed., Science Publishers Inc., Ann Arbor, 1980, 3.

Bagander, L.E. and Niemisto, L., An evaluation of the use of redox measurements for characterizing recent sediments, *Estuarine Coastal Mar. Sci.*, 6, 127, 1976.

Bates, R.G., The modern meaning of pH, *CRC Crit. Rev. Anal Chem.*, 10, 247, 1981.

Berner, R.A., Electrode studies of hydrogen sulphide in marine sediments, *Geochim. Cosmochim. Acta*, 27, 563, 1963.

Berner, R.A., A new geochemical classification of sedimentary environments, *J. Sed. Petrol.*, 51, 359, 1981.

Breteler, R.J., Scott, K.J., and Shepherd, S.P., Application of a new sediment toxicity test using the marine amphipod *Ampelisea abdita* to San Francisco Bay sediments, in *Aquatic Toxicology and Hazard Assessment*, Vol. 12, Cowgill, U.M. and Williams, L.R., Eds., ASTM, Philadelphia, 1989, 304.

Carignan, R., Rapin, F., and Tessier, A., Sediment pore water sampling for metal analysis: a comparison of techniques, *Geochim. Cosmochim. Acta*, 49, 2493, 1985.

Chapman, P., Marine sediment toxicity tests, in *Chemical and Biological Characterization of Municipal Sludges, Sediments, Dredge Spoils, and Drilling Muds*, Lichtenberg, J.J., Winter, J.A., Weber, C.I., and Fradkin, L., Eds., ASTM Special Technical Publication (STP) 976, Philadelphia, 1988, 391.

Filipek, L.H. and Owen, R.M., Geochemical associations and grain-size partitioning of heavy metals in lacustrine sediments, *Chem. Geol.*, 26, 105, 1979.

Gupta, S.K. and Chen, K.Y., Partitioning of trace metals in selective chemical fractions of nearshore sediments, *Environ. Lett.*, 10, 129, 1975.

Håkanson, L. and Jansson, M., *Principles of Lake Sedimentology*, Springer-Verlag, Berlin, 1983, 32.

Jenne, E.A. and Luoma, S.N., Forms of trace elements in soils, sediments, and associated waters: an overview of their determination and biological availability, in *Biological Implications of Metals in the Environment*, Wildung, R.R. and Drucker, H., Eds., U.S. Energy Res. Develop. Admin. Sym. Ser. 42, 1977, 110.

Kemp, A.L.W. and Lewis, C.F.M., A preliminary investigation of chlorophyll degradation products in the sediments of Lakes Erie and Ontario, in *Proc. 11th Conf. Great Lakes Res.*, Inter. Assoc. Great Lakes Res., Ann Arbor, 206, 1968.

Kemp, A.L.W., Savile, H.A., Gray, C.B., and Mudrochova, A., A simple corer and a method for sampling the mud-water interface, *Limnol. Oceanogr.*, 16, 689, 1971.

Landrum, P.F., Faust, W.R., and Eadie, B.J., Bioavailability and toxicity of a mixture of sediment-associated chlorinated hydrocarbons to the amphipod *Pontopeia hoy*, in *Aquatic Toxicology and Hazard Assessment*, Vol. 12, Cowgill, U.M. and Williams, L.R., Eds., ASTM, Philadelphia, 1989, 315.

Martin, W.F., Lippit, J.M., and Prothero, T.G., *Hazardous Waste Handbook for Health and Safety*, Butterworth-Heinemann, 1992, 307.

Morris, J.C. and Stumm, W., Redox equilibria and measurements in the aquatic environment, *Adv. Chem. Ser.*, 67, 270, 1967.

Mortimer, C.H., The exchange of dissolved substances between mud and water in lakes, *J. Ecol.*, 29, 280, 1941.

Mortimer, C.H., The exchange of dissolved substances between mud and water in lakes, *J. Ecol.*, 30, 147, 1942.

Rapin, F., Tessier, A., Campbell, P.G.C., and Carignan, R., Potential artifacts in the determination of metal partitioning in sediments by a sequential extraction procedure, *Environ. Sci. Technol.*, 20, 836, 1986.

Rozanov, A.G., Pacific sediments from Japan to Mexico: some redox characteristics, in *The Dynamic Environment of the Ocean Floor*, Fanning, K.A. and Manheim, F.T., Eds., Lexington Books, 1982, 239.

Shaheen, E.I., *Technology of Environmental Pollution Control*, 2nd ed., Penwell Publishing Company, Oklahoma, 1992, 557.

Tessier, A., Campbell, P.G.C., and Bisson, M., Sequential extraction procedure for the speciation of particulate trace metals, *Anal. Chem.*, 51, 844, 1979.

Tessier, A., Campbell, P.G.C., and Bisson, M., Particulate trace metal speciation in stream sediments and relationships with grain size: implications for geochemical exploration, *J. Geochem. Explor.*, 16, 77, 1982.

Thorstenson, D.C., The concept of electron activity and its relation to redox potentials in aqueous geochemical systems, *U.S. Geological Survey Open-File 84-072*, 1984, 45.

Tinsley, I.J., *Chemical Concepts in Pollutant Behaviour*, A. Wiley — Interscience Publication, 1979, 92.

Whitfield, M., Eh as an operational parameter in estuarine studies, *Limnol. Oceanogr.*, 14, 547, 1969.

Whitfield, M., Thermodynamic limitations on the use of the platinum electrode in Eh measurements, *Limnol. Oceanogr.*, 19, 857, 1974.

Williams, J.D.H. and Pashley, A.E., Lightweight corer designed for sampling very soft sediments, *J. Fish Res. Board. Can.*, 36, 241, 1979.

Zobell, C.E., Studies on the redox potential of marine sediments, *Bull. Am. Ass. Petrol. Geol.*, 30, 1946.

CHAPTER 6

Handling, Preservation Techniques, and Storage of Sediment Samples

6.1 INTRODUCTION

Sediment samples can be collected for determining inorganic or organic substances, particle size distribution, biological tests, etc. Methods of handling, preserving, and storing the samples depend on the objective of the sample collection. Figure 6-1 represents, in a schematic way, the different stages of sediment sample handling. Each of the stages is discussed below.

Samples consisting mainly of fine-grained sediments have a relatively uniform particle-size distribution, typically particle sizes <63 μm. However, many samples can contain mixtures of fine- and coarse-grained particles. In most environmental assessment studies, one has to decide whether the coarse material should be sieved out and discarded because it contains, in most cases, relatively low concentrations of contaminants. Alternately, the entire sample could be ground to a suitable particle size, yielding a "bulk" sediment sample for analyses. Neither of these procedures can be recommended over the other, since there is still an ongoing discussion as to which technique properly represents the character of the sediment; the choice depends mainly on the study objectives. Irrespective of which procedure is used, a detailed description of sample preparation should be included in reporting the results of chemical analyses.

Not all of the handling operations described in this section are required for all samples. Depending upon the study objectives, workers should choose those operations that are necessary to properly handle the samples for their application.

6.2 CONTAINERS FOR SEDIMENT SAMPLES

Careful planning and selection of containers and utensils involved in sediment handling prior to analysis is extremely important. Containers and other equipment used in handling sediment samples after retrieval can be a

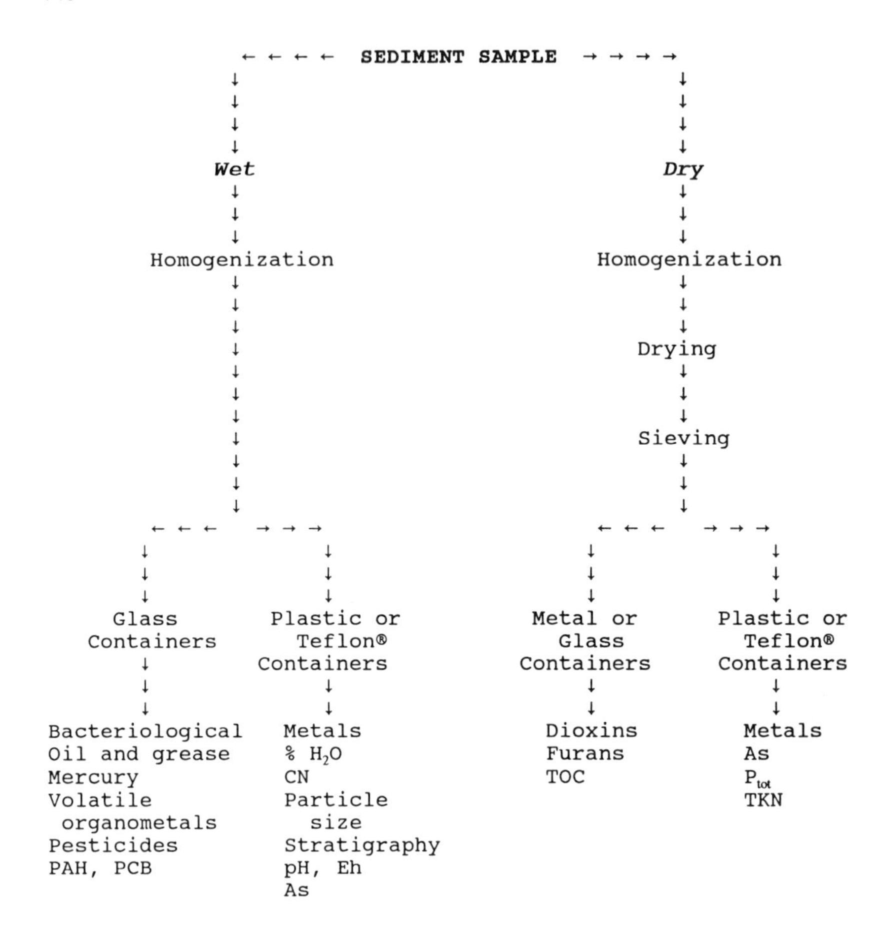

Figure 6-1 Flowchart of stages of sediment sample preparation.

significant source of contamination. Containers should neither contaminate the sample, nor promote loss of parameters of interest through adsorption on walls, etc. For example, plastics contain plasticizers that can be potential contaminants in the determination of organic compounds. Metal containers, spoons, or other equipment may contaminate samples that will be analyzed for metals and trace elements. If both organic and metal analysis are required for a given sediment sample, we recommend using a Teflon® container. Details on preventing contamination of sediment samples by different materials and procedures for cleaning the containers for sample storage are described in Sections 7.3 and 7.4, respectively.

Some common types and sizes of containers used for storage of sediment samples are shown in Figure 6-2. Synthetic polymeric plastics have found widespread application in trace metal work. Among these materials, Teflon®, polypropylene, and high-pressure polyethylene are the most desirable. Boro-

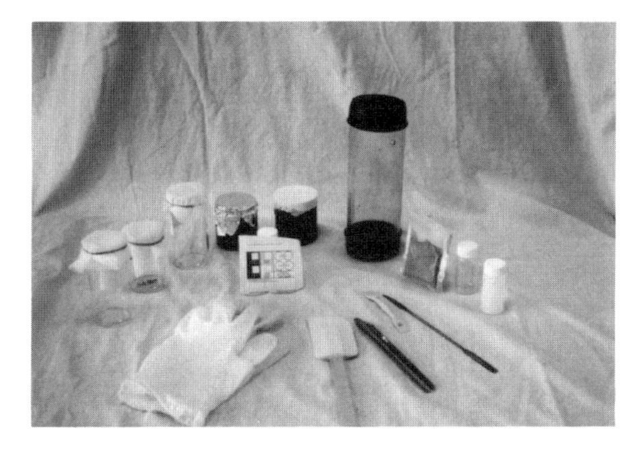

Figure 6-2 Containers and implements for sediment samples.

silicate glass has been found useful in storing samples for analysis of organic compounds, mercury, and the volatile organometallic forms of elements such as lead, bismuth, and selenium. Wide-mouth screw-capped clear and amber glass bottles in sizes from 30 ml to 4 l with Teflon®-lined caps, and wide-mouth polyethylene bottles of similar sizes with polypropylene screw caps are recommended containers that can be obtained from any scientific supply company. Precleaned containers for water and sediment samples have recently become commercially available.

The prime disadvantage of glass containers or any glass instruments is that they can easily break in the field, during shipment, or particularly when they are filled with sediments and frozen. It should be remembered that if a container with a sample needs to be frozen, it should be filled to only two-thirds of its volume. Polyethylene and Teflon® utensils used in sediment handling are usually best for sediment samples to be analyzed for inorganic components. Plastic bags of various sizes, made of polyethylene, polypropylene, or other plastic, can be used for storing wet or dry sediment samples. Sediments for biological testing can be collected, transported, and stored in plastic or glass containers (Jafvert and Wolfe, 1987; Giesy et al., 1988). Table 6-1 summarizes the recommended sample containers for collecting sediments prior to determining different parameters.

Generally, containers used for drying should be made of material resistant to corrosion and not subject to change in weight or to disintegration on repeated heating and cooling. Crucibles, dishes, and trays made of aluminum, nickel, glass, and porcelain are recommended for drying sediments in an oven. The selection of the material depends on subsequent analyses of sediment samples. Therefore, considerations in the selection of containers also apply to the choice of containers for drying.

Prior to sampling, all sediment containers should be properly labelled with a waterproof marker. Never label only the container's lid because it is very easy

Table 6-1 Sample Containers and Preservation for Different Parameters Measured in Sediments

Parameter[a]	Container[b]	Preservation	Maximum Storage	Comments
Particle size	P, G, or M	Wet, 4°C, tightly sealed	14 days	Drying, freezing, and thawing cause aggregation of particles
Stratigraphy	Core	Wet, 4°C	Several months	Preserve original consistency
Bioassays	P or G	Sieved, 4°C, dark	Processed within 2 to 7 days	Mixing and sieving recommended before testing sediment toxicity
Bacteriological	Sterile G	Wet, 4°C	Processed within 6 hours	
pH, Eh, CEC	Bucket or core	Wet undisturbed and untreated	Determined in the field	Very difficult and problematic temperature corrections
P_{tot}, TKN	G	Freeze, –20°C	1 month	If possible, analyze in 24 hours
TOC	P or G	Freeze, –20°C	6 months, dark	Carbonates and bicarbonates can interfere
Oil and grease	M or G	Wet, 4°C	1 day	Wet sample can be stored for up to 1 month at –10°C with 1–2 ml concentrated H_2SO_4 per 80 g
Metals	P or T	Dry (60°C) or freeze (–20°C)	6 months	If samples are not analyzed within 48 hours, freeze dried –20°C up to 6 months
Mercury	G or T	Freeze, –20°C	1 month	Mercury analysis is performed with wet samples
Volatile organics	G vials with Teflon® septums	Freeze, –20°C	1 month	No preservatives should be added Possible loss of some compounds
Cyanides	P	Freeze, –20°C	Up to 1 month	Sulfide interfere colorimetry
Pesticides and PCB	M or G covered with Al foil	Freeze, –20°C, dark	7 days until extraction	If samples are not analyzed within 48 hours, freeze dried –20°C up to 6 months

[a] CEC = cation exchange capacity; P_{tot} = total phosphorus; TKN = total Kjeldahl nitrogen; TOC = total organic carbon.

[b] P = polyethylene or polypropylene; G = glass; T = Teflon®; M = metal.

to mix up the lids on the sample containers during sample handling in the field. Labels should contain the following information:

- Site and sample identification
- Data and time of collection
- Sediment use
- Preservative used
- Name of collector

We highly recommend carring extra containers on a sampling expedition, exposing them to the same conditions as those actually used for samples. These containers serve as field blanks.

In considering sampling, handling, and long-term storage of a variety of environmental specimens including sediments (prior to analyzing components, many of which may be present at trace levels), Luepke (1979a) summarized the recommendations for sample handling as follows:

- Minimize interaction between samples, containers, and utensils used in sample handling.
- Minimize interaction between samples and external environment.
- Test any material which is in contact with the samples.
- Treat sample containers with the same precautions as the samples.
- Wash sample containers and all utensils for sample handling with appropriate cleaners.
- Run a proper analytical blank specific to the character and analysis of every sample.

6.3 HANDLING SAMPLES FOR TESTS AND ANALYSES ON WET SEDIMENTS

Figure 6-3 outlines the possible operations that are usually required on wet sediments. Removal of large debris and pebbles facilitates the homogenization of samples to be subsampled for multiple biological and/or chemical tests.

6.3.1 Particle Size Distribution, Geotechnical Tests, and Sediment Stratigraphy

Particle size distribution analyses should be carried out on wet sediments. Samples for the analyses should be stored at 4°C and never frozen. Tightly sealed plastic bags, glass jars, or other containers can be used to store samples prior to the particle size analyses (Table 6-1). Sediments with a high iron content should be stored in airtight containers to avoid precipitation of iron oxides on particle surfaces, and should be analyzed as soon as possible after collection. Drying, freezing, and thawing of the sediments can cause irreversible aggregation of particles and should be avoided. Size analyses of fine-

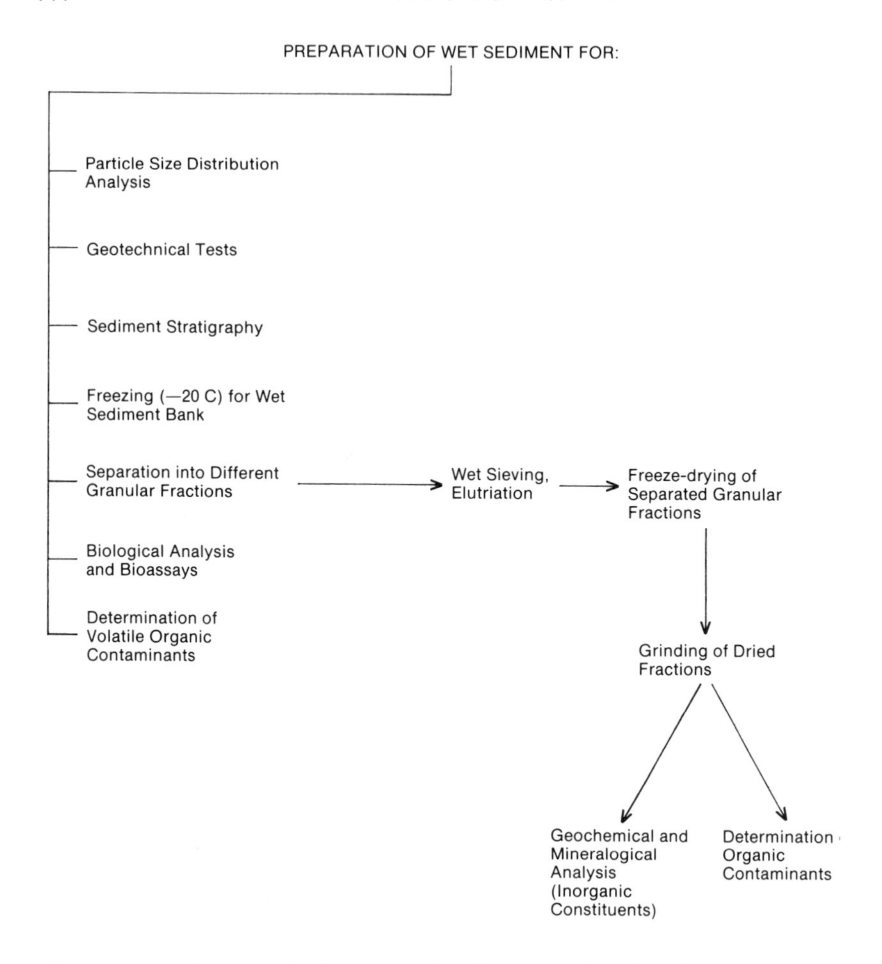

Figure 6-3 Handling samples for tests and analyses on wet sediments.

grained sediments should be carried out only on completely dispersed samples that have been treated for the removal of organic matter, and carbonate and iron coatings.

Sediment samples collected for stratigraphic or geotechnical studies can be stored at 4°C in a humidity-controlled room without any large changes in sediment properties for several months (Table 6-1). Long cores, such as those collected by piston coring, can be cut into lengths suitable for storage, sectioned longitudinally, described, labelled, wrapped to preserve their original consistency, and stored in a refrigerated room (Figure 6-4).

6.3.2 Freezing of Wet Sediments

Freezing has long been an acceptable preservation method for sediments collected for determining organic and inorganic constituents. It has been widely used for sediments and biological samples (Environment Canada, 1979; Water Quality National Laboratory, 1985). Freezing, but not thawing and refreezing, is the preferred method of keeping samples for organic analyses. However, it should not be used for particle analyses and toxicity tests. Luepke (1979b) reported that rapid and deep freezing can best maintain sample integrity and thus enable the investigation for contaminants concentrations. The lower the temperature of deep freezing, the better: a temperature of –80°C is the suggested maximum. Gills and Rook (1979) found little variation in concentrations of selected trace elements on samples for biological tests that were stored for one year at –80°C compared with fresh samples.

Rutledge and Fleeger (1988) found a dramatic effect on the vertical profile of meiobenthos when sediment cores were preserved by fast freezing. Freezing the whole cores by any method can cause some distortion of the sediment's vertical profile. However, the size of the corer had no discernible effect on core distortion. Distortion was strongly related to the temperature at which the cores were frozen, with fast freezing tending to cause a greater disruption of the natural sediment profile.

Freezing sediments requires a considerable expense in purchasing and maintaining the equipment. Generally, the lower the required temperature, the greater the cost incurred. The size of the sample set(s) to be preserved and the length of time for which they must be stored enter into the cost estimate. Also, the possibility of catastrophic freezer failure and the necessity of warning and/or backup systems must be considered. Good freezer control is required to minimize temperature fluctuations.

Wet Sieving

Wet sieving is particularly useful for processing fine-grained sediments. Wet sieving of a small quantity of sediment is generally carried out manually

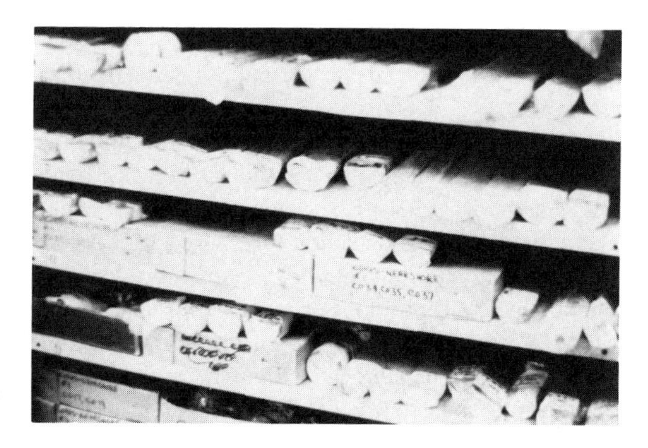

Figure 6-4 Storage of cores for sediment stratigraphy.

and requires one or more sieves and two types of containers, such as buckets, dishes, beakers, bowls, and jars. When a dry sediment is to be wet-sieved, the sample is placed in a container (usually a metal or plastic bucket or large dish), clean water is added to cover the sample, and a minimum of two hours is allowed for soaking. A portion of soaked sample, which may be weighed if desired, is then placed on the standard sieve with the appropriate opening, i.e., 0.5 mm (35 mesh). The sieve is placed in a dish filled with water and gently swirled so that particles smaller than the selected sieve size are washed through the sieve into the dish. In a method described by ASTM (1986), all material in the container is transferred to the sieve and washed with the smallest volume of running water. To facilitate passing through the sieve, a nylon brush can be used. Particles retained in the sieve (the coarse fraction) are examined and, if they are of interest to the study, retained and redried. The sediment washed through the sieve (the fines) is redried and weighed. After use, all sieves should be thoroughly cleaned under running water, with special care to remove the material caught by the screen.

6.3.3 Separation into Granular Fractions

Sediments are often used to monitor metal pollution in natural waters. Sediment grain size, which may strongly affect the results of such investigations, is often neglected. Examples of approaches to separating sediments into various particle size fractions and methods used for separation are discussed later in this section.

Studies on the correlation between metals and sediment particle size fractions (Ackermann, 1980; de Groot and Zschuppe, 1981; de Groot et al., 1982) suggest that fine-grained sediments usually contain greater concentrations of metals, and that the main portion of many metals is incorporated in the silt and clay size fractions (<63 μm particle size). One of several methods to investigate the significance of the granular composition in the concentration of metals or other contaminants in sediments is based on the isolation of individual granular fractions and determination of metal concentrations in each fraction. Ackermann et al. (1983) reviewed several studies and suggested that the choice of the <60 or <63 μm size division was based on the traditional definition of the silt and sand boundary. A major advantage in using the <63 μm size fraction proposed by Förstner and Salomons (1980) is the greater concentrations of trace elements (i.e., above the analytical detection limit) associated with silt/clay particles. In addition, many studies have been carried out using this sediment fraction to compare the concentrations of metals or organic compounds over a larger area. More information on sampling of fine-grained sediments is given in Section 2.4. In addition, the silt/clay sediment size fraction is close to that of material carried in suspension, which is important in studies of sediment resuspension and transport.

Wet sieving has been used to separate different particle size fractions in sediments. It should be noted that wet sieving involves resuspension of sediments in water and may change their original size distribution. Furthermore, because the water used for wet sieving usually does not have the *in situ* ionic composition, resuspension may even break particles that were originally agglomerated (Adams et al., 1980). This effect may be reduced by using water collected at the sediment sampling site.

In their study, Wilberg and Hunter (1979) used water from the river where bottom sediments were collected for wet sieving. Three 1-l portions of water were passed through a continuous flow centrifuge to remove suspended matter prior to wet sieving of the sediment through six stainless-steel sieves of sizes 2,000, 1,000, 420, 250, 125, and 63 μm. The remaining 3-l suspension of particles <63 μm was fractionated by sedimentation-decantation and successive centrifugation using a continuous flow centrifuge. The silt/clay suspension was not treated prior to separation in order to retain the natural aggregate state of the particles. Wall et al. (1978) suggested that during separation of the silt/clay fraction from the bulk sediment using chemical dispersants, mechanical stirring, and subsequent sieving, considerable quantities of smaller particles adhere to the larger ones. Similar problems can be encountered during dry or wet sieving due to improper use of the ultrasonic treatment (Ackermann, 1980).

Ackermann et al. (1983) used plastic sieves in a beaker placed in an ultrasonic bath with 70 to 100 ml of distilled water to separate the <60 μm and <20 μm particle size fractions. The grain size distribution varied only slightly (1 to 4%) according to the duration of the sieving process. Treating several samples at the same time, the separation of a sediment sample into three fractions (<20 μm, 20 to 60 μm, and 60 to 200 μm) required approximately 15 to 20 minutes per sample.

Another system for separating particles into size ranges is based on the principle of elutriation. A wet sediment sample is separated into specific size fractions by a process that depends upon the forces present in a moving fluid. The Cyclosizer is a commercially available instrument (Warman International Ltd., 1981) that separates particles according to their relative size and density, in a series of hydraulic cyclones where the centrifugal force produces the elutriating action (Figure 6-5). This unit is capable of separating silt-sized particles into six standard fractions: >44 μm, 33 to ≤44 μm, 23 to ≤33 μm, 15 to ≤23 μm, 11 to ≤15 μm, and <11 μm. These size ranges are nominial and depend upon standard conditions of water flow rate, water temperature, particle density, and time of separation. Deviations from standard conditions will result in deviations in the size cutoffs. The utility of this system for silt-sized particles has been demonstrated by Mudroch and Duncan (1986) on sediments from the Niagara River, and by Stone and Mudroch (1989) for Lake Erie sediments. The Cyclosizer enables the separation of large amounts of a sample in a batch

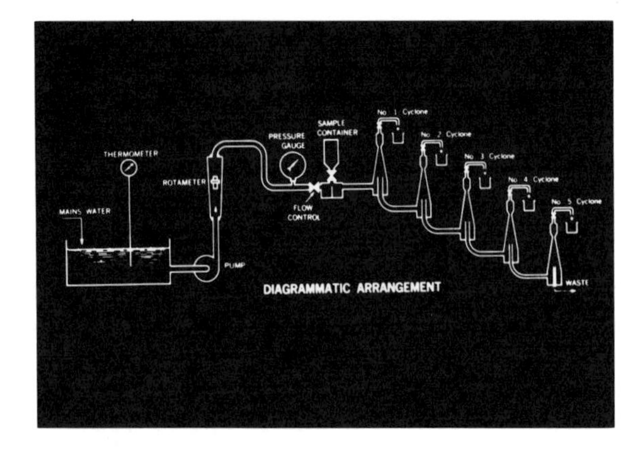

a

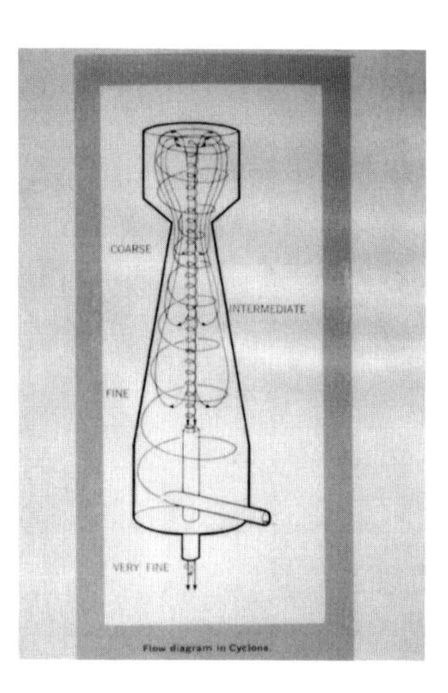

b

Figure 6-5 Cyclosizer for separation of sediments into different size fractions: (a) schematic diagram of the complete process, (b) flow diagram in cyclone.

system for subsequent chemical analysis. This method provides the separation ability that was once only practical for coarser samples by wet sieving.

Umlauf and Bierl (1987) used an elutriation procedure to study the partitioning of several organic pollutants in sediments and suspended solids in a river. The elutriation involved suspending the sediment sample in an upward flow of the dispersing medium (0.01 M sodium pyrophosphate). Particles that settled at velocities greater than the upward velocity of the medium remained in the separation chamber, and those settling at lower velocities were carried upward to the next separation chamber. The upward flow can be calibrated by the calculation of the settling velocity using Stokes' law to achieve separation into required particle size fractions.

In a study of mercury distribution in an estuary, Cranston (1976) separated sediment samples that were previously dried at 40°C into size fractions by sieving through a series of sieves (>1,000 μm, 500 to ≤1,000 μm, 250 to ≤500 μm, 125 to ≤250 μm, 63 to ≤125 μm, and <63 μm). The <63-μm-size fraction was further subdivided into 16 to ≤63 μm and <16 μm fractions using an Atterburg sedimentation column. The 16 to ≤63 μm fraction was collected from the bottom of the tube, and the finest fraction collected by centrifuging the decanted suspension from the sedimentation column. For this study, dry sediment samples were lightly pulverized using an agate mortar and pestle and soaked in water for about 3 hours before wet sieving and elutriation.

6.3.4 Handling Samples for Biological Analyses and Bioassays

We recommend processing samples collected for the investigation of benthic organisms in the field by wet sieving through sieves of required size. Mixing and sieving are often required before testing sediment for toxicity. Although it is generally accepted that sediment sieving can alter the chemical character of collected sediments, it has been used in many studies of sediment toxicity. For example, sieving contaminated sediments through a 250-μm mesh decreased concentrations of PCBs and polynuclear aromatic hydrocarbons (PAHs) as much as four-fold (Day et al., 1992). Sediment particles smaller than 1 mm are usually retained for testing (Swindoll and Applehaus, 1987; Landrum and Poore, 1988; Landrum et al., 1992). Removal of large debris and pebbles facilitates homogenization of samples to be subsampled for multiple biological and/or chemical tests. Counting benthic organisms is also easier on samples cleaned of debris and pebbles. Sieving is carried out to remove endemic species that can interfere in the biological testing of sediments (Day, 1994). If, for any reason, the samples cannot be processed in the field, they should be stored at 4°C in the dark and processed in the laboratory as soon as possible, preferably within 48 hours. Biological tests should be conducted within two to seven days (Swartz et al., 1985; Burton and Stemmer, 1987; Anderson et al., 1987; Klump et al., 1987; Breteler et al., 1989). Storage of sediments collected for bioassays is described in Section 6.6.

Wet sieving or pressure-sieving of sediments were used in biological tests conducted by Burton et al., 1989, Giesy et al., 1990, Stemmer et al., 1990, Mudroch and MacKnight, 1994. Details of wet sieving are described in Section 6.4.2. Pressure sieving is carried out by pressing sediment particles through a sieve with a desired mesh size using mechanical, piston-type equipment or a flat-surfaced, manually operated tool.

In some studies, samples are collected for spiking the sediments with different elements and compounds. The spiked sediments are further used in bioassays to determine the effects of the element or compound introduced to the sediments on different species. Collected sediments have sometimes been air-dried prior to spiking (Foster et al.,1987; Keilty et al., 1988; Landrum and Faust, 1991). Currently, however, most acceptable sediment spiking techniques use wet sediments (Francis et al., 1984; Birge et al., 1987; Suedel et al., 1993). Therefore, collected sediments have to be stored and prepared for spiking according to the study design and protocol describing the procedures for sediment spiking relevant to the study. During spiking, a solution of element(s) and/or compound(s) can be added to a sediment slurry, or can be coated on the wall of a flask with the subsequent addition of the sediment slurry. The solution can also be added to the water overlying the sediments to allow adsorption of the element or compound into sediment particles without mixing (Tsushimoto et al., 1982; Lay et al., 1984; Stephenson and Kane, 1984; Crossland and Wolff, 1985; O'Neill et al., 1985; Gerould and Gloss, 1986; Pritchard et al., 1986).

Spiked sediments need to be completely homogenized before they can be used in bioassays. Different mixing methods have been reported for the homogenization of spiked sediments, such as the rolling mill technique (Ditsworth et al., 1990; Swartz et al., 1990; DeWitt et al., 1992) and use of shakers (Stemmer et al., 1990). Proper homogenization is necessary, particularly when spiked sediments are to be subsampled for bioassays and for chemical analyses. We recommend conducting chemical analyses to ensure the homogeneity of spiked sediments. Further, mixing spiked sediments should be time-limited and carried out at a low temperature, such as 4°C, to prevent changes in the microbiological and physico-chemical character of the sediments. The homogeneity of sediments spiked and mixed by different techniques was investigated and expressed as a coefficient of variation in concentrations of the spiked element or compound in subsamples collected from homogenized sediments (Ditsworth et al., 1990; Landrum and Faust, 1991; Landrum et al., 1992; Suedel et al., 1993). The results indicated that tried techniques produced a relatively homogeneous mixture of spiked sediments.

6.4 HANDLING SAMPLES FOR TESTS AND ANALYSES ON DRY SEDIMENTS

Sediments, particularly the topmost 10 cm, typically contain up to 95% water. To permit the comparison of obtained data, sediments are dried and analyses are carried out on dry material, or a subsample is taken for drying to determine the water content while the analyses are performed on the wet sediment. In either case, results of analyses are usually presented on a "dry weight" basis. Further, some analytical methods, such as X-ray fluorescence spectrometry, require dry material ground to a certain particle size. Therefore, a list of analytical methods and requirements for dry sample preparation must be prepared prior to the collection of the samples.

Handling operations for dry sediment are shown in Figure 6-6. The operations required for preparating dry sediment samples — such as drying, sieving, grinding, mixing, and homogenizing — are described in detail below.

6.4.1 Drying

Three types of drying are commonly used to prepare solid samples prior to analysis: air-drying, oven-drying, and freeze-drying.

Air-Drying

Although time consuming, air-drying is commonly used in soil science (McKeague, 1978) and in sedimentology (Folk, 1974). It is only rarely used for the preparation of sediments for pollution studies. Air-drying may generate undesirable changes in sediment properties and increases the risks of contamination. For example, changes in metal availability and complexation for samples that were air-dried was documented by Kersten and Förstner (1987). On the other hand, air-drying has been used in some studies to avoid the loss of components, such as mercury, that are volatile at temperatures greater than 50–60°C (Förstner and Salomons, 1980).

The difficulty is to achieve thorough drying to constant weight as specified by the ASTM method D421-58 (ASTM, 1969a). S.D. MacKnight has tested the air-drying of marine sediments (personal communication). Organic matter-rich, fine-grained sediments typical of harbors were found to take three to five days of air-drying to achieve a constant weight. A typical drying period of 24 hours was found to remove only 40 to 60% of the water content. Slow air-drying was attributed to hygroscopic salts and organic material in the samples. Drying sediments from areas adjacent to pulp mills was found to be even more difficult due to wood particles and other organic material in the samples.

Sediments collected for plant bioassays (Folsom et al., 1981; Van Driel et al., 1985; Mudroch and Painter, 1987) are usually partially air-dried at 20 to 40°C with relative humidity at 20 to 60%. Depending on the quantity of material, air-drying is carried out in a fume hood (small samples), air-drying cabinet with air circulation, or in sheltered ventilated rooms (large samples). However, because of the possibility of air contamination of sediment samples by dust, air-drying is not recommended for the accurate determination of inorganic and organic constituents in the sediments. Similarly, because of biological activity during the drying period, air-drying is not recommended for sediment samples to be used for biological tests. For chemical analyses, where

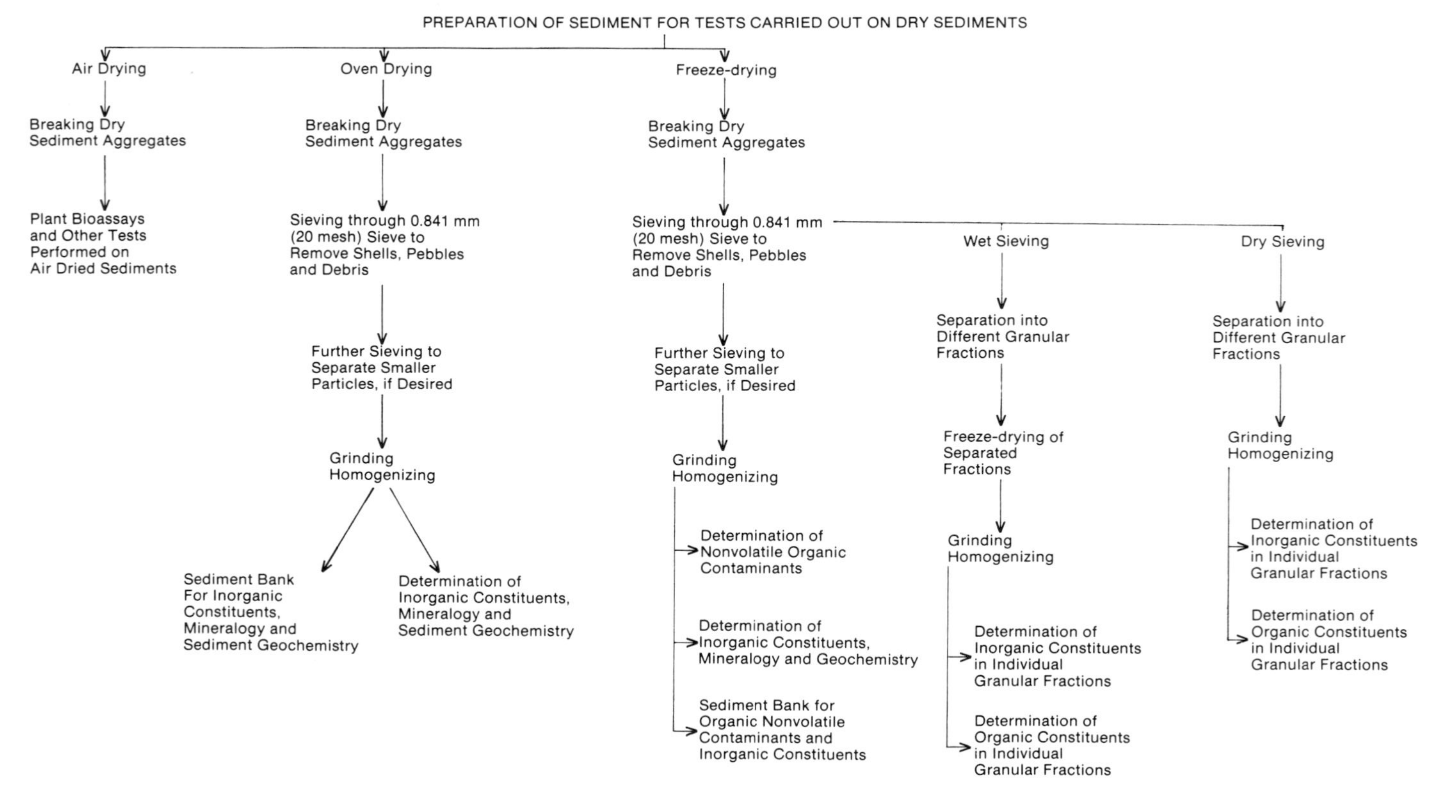

Figure 6-6 Handling samples for tests and analyses on dry sediments.

preservation is required in addition to the drying, this method is not suitable because microbial degradation, oxidation, and other processes that can alter the sample are not halted.

Oven-Drying

Oven-drying of sediments is usually carried out on samples collected for determining inorganic components, such as major and trace elements. However, oven-drying is not suitable for grain size determination, since wet fine-grained sediments become hard-to-break aggregates (Folk, 1974). Oven-drying is not acceptable for sediments containing any volatile or oxidizable components, whether they be organic or inorganic (Luepke, 1979a), and may contribute to the alteration of even nonvolatile organics. For instance, dehydration of aliphatic alcohols can create unsaturation, and heat, in the presence of clay catalysts, can promote pyrolytic reactions.

Geological materials in general, and sediments in particular, on heating to 100 to 110°C in an oven, release most of their hygroscopic water, which is water held by surface forces such as adsorption and capillarity. The amount of hygroscopic water is related to the physical properties and mineralogical composition of sediments. Sediment interstitial water is also evaporated during the drying procedure. The ASTM D22-63T method (ASTM, 1964) covers the laboratory determination of the moisture content of soil at a temperature of 110 ± 5°C. This is similar to the method used in geology (Maxwell, 1968) and can be used for determining moisture in sediments. Lower heating temperatures (less than 60°C) are essential when preparing sediment for the determination of volatile trace elements, such as mercury (Luepke,1979b). The effects of various drying temperatures on the determination of mercury in sediments were reviewed by de Groot and Zschuppe (1981). They concluded that mercury should be determined in sediments that have preferably either been air-dried or oven-dried at 40°C. Drying ovens are available from companies supplying laboratory equipment in a variety of models with capacities from 20 to 1,600 l, temperature ranges from 40 to 250°C, air circulation, and accurate control of temperature. An alternative drying procedure for these analyses is freeze-drying and is described below.

Freeze-Drying

In the freeze-drying process, water in the frozen or solid state is sublimated and removed from the material as a vapor. Freeze-drying (also called lyophilization) can be used for drying sediments collected for the determination of most organic pollutants as well as for analyses of inorganic components, such as major and trace elements. Certain organic components are more susceptible to volatilization. For example, loss of lighter chlorobenzenes (di- and trichlorobenzene) was observed during testing of the effects of freeze-drying on the integrity of sediment samples from the Great Lakes (Bourbonniere et al., 1986). Fox et al. (1991) observed that inadequate temperature control and excessive total exposure time during freeze-drying produced large and highly variable losses of naphthalene and much smaller losses of higher molecular weight PAHs. Contaminated sediment samples from Hamilton Harbour, Lake Ontario, were sectioned at 2-cm intervals immediately after collection. Samples to be freeze-dried were thawed, blended, subsampled, and refrozen immediately prior to freeze-drying. The samples were freeze-dried for approximately one week. Similarly, samples to be wet-extracted were thawed, blended, and subsampled. Excess water was removed from the wet extraction subsamples by decanting and absorption on a pad of glass-fiber filters. Mean losses of naphthalene on freeze-dried sediments was 96%, losses of pyrene were 42%, and benzoperylene 14% (Table 6-2). The trend in these decreasing losses is most readily explained by the increase in molecular weight with concomitant decreases in vapor pressure. Although the exposure time of the sample to freeze-drying conditions was arguably excessive, the procedure used was typical of a busy laboratory processing large numbers of samples, and also gives a good indication of how extreme the losses may be (Fox et al., 1991).

Some inorganic constituents, such as mercury and iodine, can also be lost (Pillay et al., 1971; LaFleur, 1973; Harrison and Lafleur, 1975). De Groot and Zschuppe (1981) reviewed reports on losses of mercury by freeze-drying sediments. They found that no detectable loss of mercury following freeze-drying was reported by some authors, but one investigator reported that results from freeze-dried samples were 23% lower than those from air-dried samples. During drying (freeze-drying and oven-drying) of large samples, a cake may

Table 6-2 Concentrations of Selected PAHs (µg/g) in Sediments from a Coal-Tar-Contaminated Site in Hamilton Harbor Extracted Wet and After Freeze-Drying

	Wet sediment	Freeze-dried sediment	Loss (%)
Naphthalene	7041	61	99
Acenaphthylene	5	4	20
Acenaphtene	22	2	91
Fluorene	19	4	79
Phenanthrene	71	41	42
Anthracene	24	3	87
Fluoranthene	65	40	38
Pyrene	47	27	42
Benzoanthracene	25	15	40
Chrysene	23	18	22
Benzofluoranthracene	41	22	46
Benzo(a)pyrene	22	13	41
Benzo(e)pyrene	13	11	15
Pyrelene	6	4	33
Indenopyrene	14	11	21
Dibenzoanthracene	4	3	25
Benzoperylene	14	12	14

Modified from Fox et al. (1991).

form with wet sediment inside. We recommend gently hand-grinding the cake and continuing the drying process.

The following are the principal advantages of freeze-drying for sediments:

- the low temperatures avoid chemical changes in labile components;
- the loss of volatile constituents, including certain organic compounds, is minimized (Bourbonniere et al., 1986);
- most particles of dried sediments remain dispersed;
- the aggregation of the particles is minimized;
- sterility is maintained;
- oxidation of various minerals or organic compounds is minimized or eliminated.

The price of freeze-drying equipment — together with the need for special bottles and vials — and the maintenance cost are significantly greater than the price and maintenance cost of drying ovens. At the National Water Research Institute, Environment Canada, Burlington, Ontario, the procedures for freeze-drying can vary depending on which units are available and the type of analyses to be performed on freeze-dried samples. Several commercially available freeze-driers are operated according to the recommendations appearing in the manufacturers' manuals.

Pre-freezing of samples is common when freeze-drying is anticipated. This is carried out mainly to avoid "bumping" or spattering the sample when evacuating the freeze-drier chamber, and has the added advantage of offering intermittent preservation. LaFleur (1973) compared room temperature storage, pre-freezing at –20°C, and at –196°C, i.e., using liquid nitrogen prior for freeze-drying muscle and liver tissue. No difference was found among these methods for retention of methylmercury chloride. Alternatively, unfrozen sediment samples can be loaded directly into the freeze-drier in suitable containers. Many commercial units, such as that in Figure 6-7, have the capability of a freezing cycle prior to the drying cycle.

The essential steps in freeze-drying sediments in most commercial freeze-driers are:

- Replace the caps on vials, jars, or bottles containing samples with filter paper and special tops. If samples are freeze-dried in plastic bags, open the bags and cover the top with a filter paper or a tissue that would not contaminate the sample. Secure the filter paper or tissue on top of the containers using masking tape or rubber bands. The filter paper allows water to escape while retaining the sediment particles during freeze-drying and particularly during release of the vacuum in the drying chamber. Special freeze-dry flasks are commercially available. These flasks are attached to the valves mounted on the ports outside the drying chamber by an adaptor. Samples collected in small containers can be placed in the special freeze-dry flasks. Filter papers are also available that fit into the top of these flasks to prevent loss or contamination of freeze-dried samples.

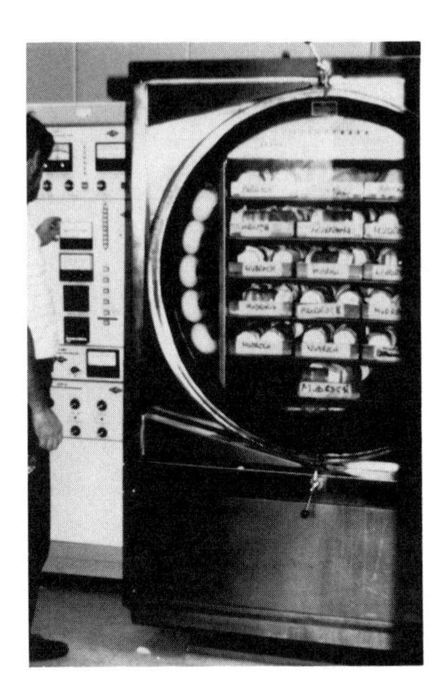

Figure 6-7 Freeze drier.

- Freeze the samples for approximately 18 hours at –20°C, directly in the freeze-drier on trays in the drying chamber. Omit this step when using pre-frozen samples.
- Turn on the condenser of the freeze-drier and cool to –40° to –50°C. When sufficiently cold, apply vacuum using the high vacuum pump required by most units. The normal operating range is 0.010 to 0.050 Torr depending upon the surface area available for sublimation of water. In some units, the trays in the drying chamber can be heated to speed up the process. Such heating should be turned on only after maximum available vacuum is reached.
- The time required for complete drying in freeze-driers ranges between 24 hours and 14 days, depending on sample quantity, surface area, sediment type, and water content. For example, a 100-g fine-grained sediment sample needs about three days, and a 500-g sample of similar consistency about seven days of freeze drying. The capacity of a freeze-drying unit is usually expressed as the mass of water that can be frozen onto the condenser. The rate of drying levels off as ice builds up on the condenser. A good rule of thumb is to load the unit to about 80% of the condenser capacity.
- Check the condition of the samples after the estimated drying time by:
  (1) Slowly releasing vacuum and removing containers from the drying chamber.
  (2) Removing special tops and filter paper.
  (3) Examining the entire sample with a spatula or knife to see if it is dry, paying special attention to the center of the sample, which usually needs the longest time to dry.

If the samples are at ambient temperature and a cold spot can be felt, especially on the bottom of the container, then ice remains. If a portion of the samples is still wet, continue freeze-drying.

- When the samples are dry, mount the original caps and lids on the containers and store the samples for further processing.
- After the dry samples are removed from the drying chamber, check if the freeze-drier needs cleaning and defrosting, and control the level and quality of oil in the vacuum pump. Maintenance of the equipment should be carried out following the instructions in the manual supplied by the company.

6.4.2 Sieving

Sieving is an effective and economical process for dividing sediment samples into fractions containing particles of more or less the same size. It is common practice for soil scientists to work soil samples, which are very similar to sediments, through a 2-mm round-holed sieve (10/20 mesh) using a rubber stopper or a rubber pestle. For routine soil testing, the fraction coarser than 2 mm is discarded. In research studies, the material retained on the 2-mm round-holed sieve is examined and described, and then either discarded or, if of interest, preserved, dried, weighed, and analyzed.

If sieving of sediment samples is to be carried out, due consideration should be given to the analyses and tests planned, as discussed in the introduction to this chapter. There are three possible scenarios:

1. The whole bulk sample is ground with no sieving, or alternate grinding and sieving in a disc mill, and analyses are carried out on an homogeneous subsample.
2. The whole sediment sample is passed through a sieve of the desired mesh size, e.g., 2 mm, 63 μm (10, 250 mesh). The oversize portion is discarded while the portion that passes is used for analyses.
3. The whole sediment sample is separated into specific size fractions by dry or wet sieving and elutriation. The resulting size fractions are analyzed individually in their original state or ground to an appropriate grain size for analyses. The composition of the sediment can vary considerably for different particle sizes; therefore, thorough mixing of the material after sieving is essential.

Table 6-3 shows the classification of particles of a different size together with designated sieves for separating. Sediments are usually classified simply as gravel (>2 mm), sand (63 μm to <2 mm), silt (2 μm to <63 μm), and clay (<2 μm). These definitions are based on arbitrary cuts of median diameter between clay and silt, silt and sand, and sand and gravel; there are several classification systems (Folk, 1974). Particle size can be expressed on a millimeter scale, micrometer scale, or phi-scale. The mathematical definition of the latter is:

Table 6-3 Wentworth Size Classes, Grain Size Scale, and Sieve Numbers

Wentworth size class	Phi	Metric mm	Metric µm	U.S. Standard Sieves mm	U.S. Standard Sieves µm	No.	inch
Cobble gravel	–8	256					
	–6	64		64			$2^1/_2$
Pebble gravel	–5	32		32			$1^1/_2$
				25			1
				19			3/4
	–4	16		16			5/8
				12.5			1/2
				9.5			3/8
	–3	8		8			5/16
				6.3			1/4
				5.6		$3^1/_2$	
				4.75		4	
Granule gravel	–2	4		4		5	
				3.35		6	
				2.80		7	
				2.36		8	
Very coarse sand	–1	2		2		10	
				1.70		12	
				1.40		14	
				1.18		16	
Coarse sand	0	1		1.00		18	
					850	20	
					710	25	
					600	30	
Medium sand	1	.50			500	35	
					425	40	
					355	45	
					300	50	
Fine sand	2	.25			250	60	
					212	70	
					180	80	
					150	100	
Very fine sand	3	.125			125	120	
					106	140	
					90	170	
					75	200	
Coarse silt	4	.063			63	230	
					53	270	
					45	325	
					38	400	
	5		31				
Medium silt	6		15.6				
Fine silt	7		7.8				
Very fine silt	8		3.9				
Clay	9		2.0				
	10		.98				
	11		.49				
	12		.24				
	13		.12				
	14		.06				

$$\text{phi} = -\log_2 (d)$$

where d = particle size diameter in mm.

Dry Sieving

Dry sieving is used for separating coarse samples into several size fractions, and quite often for separating coarse from fine fractions. However, when sediments contain very fine particles, particularly clays, fine sieves become clogged, thus impeding proper sieving of the sediment. In such cases, wet sieving is employed. A combination of dry and wet sieving is often effective and is recommended: dry sieving for coarser particles and wet sieving for particles passing easily through the finest sieve. Material smaller than the finest sieve, usually 44 and 37 μm (325/400 mesh), is determined by difference, although these particles can be recovered and weighed.

Hand sieving of dry material is common. The sieve is placed on a pan, an appropriate quantity of the sediment is placed on the sieve, the sieve is covered with a lid and shaken with a rotating intermittent tapping action until separation is complete. After use, all sieves should be cleaned either with a soft paint brush or under running water and air- or oven-dried. Care must be taken to avoid contamination by the material of the sieve. If trace metals are to be determined, plastic woven sieves should be used. After use, the plastic sieve is either discarded or cleaned and reused.

Types of Sieves

U.S. standard sieves consist of a set of fine and coarse 20-cm (8-in) diameter sieves made in accordance with specifications outlined in ASTM E-11-61 (ASTM, 1969b), approved U.S. standard Z23.1, AASHO M92, and federal specifications RR-S-3668. The U.S. standard sieve designations (in mm and μm) correspond to test sieve aperture values recommended by the International Standards Organization. The U.S. series alternate sieve designations (by number) are the approximate number of openings per linear inch.

Sieves are circular frames, made of stainless steel or brass, the standard size being 20 cm in diameter, either 5 or 2.5 cm high with a wire cloth carefully soldered to the frame. Sieve covers and receivers (catch pans) made of brass or stainless steel are essential parts of any sieving equipment. For special purposes, both large- and small-diameter frames, covers, and catch pans are commercially available.

Available screening surfaces are woven wire cloth, plastic woven polymer filter screens, punched plate, and bar screens. However, only the woven wire cloth and plastic woven polymer filter screen surfaces are suitable for laboratory sediment handling. The opening, wire diameter, and open area are to be carefully considered when selecting a screening machine and sieves. Woven

Figure 6-8 Mechanical shakers for sediment sieving.

wire cloth has by far the greater selection as to screen openings from 100 mm to 37 μm (4 in to 400 mesh), wire diameter (6.3 mm to 37 μm), and percentage of open area. Woven wire screens are made mainly of stainless steel and brass, but can be made from other metals and alloys when required.

Synthetic polymer woven screens are available in four materials: polyethylene, polypropylene, nylon, and polytetrafluoroethylene (Teflon® or PTFE), with a range of sieve openings from 1 μm to 1.24 mm. They are ideal for screening solutions or suspensions containing particles to be analyzed for trace elements. For small volumes of sediment, a polyethylene microsieve set 51 mm I.D. × 29 mm supplied with all components needed to build a stack of four sieves is useful.

A number of laboratory electromagnetic and mechanical sieve shakers are on the market that can automatically carry out dry or wet sieving with accuracy and reliability (Figure 6-8). They are designed to hold and handle from one to thirteen standard 200-mm diameter sieves with different openings. Any laboratory planning to purchase a sieve shaker should consider these factors: ease of operation, noise level, space requirement, number of sieves shaken at one time, and suitability for dry and wet sieving.

A discussion of screening for industrial use of sediments is outside the scope of this book; the reader is referred to reference books such as *Chemical Engineers' Handbook* (Perry, 1963) and *Handbook of Mineral Dressing* (Taggart, 1945).

6.4.3 Grinding

Geological samples — rocks from about 3 to 15 cm containing minerals of different hardness — are reduced to powder prior to analysis in three steps: crushing, pulverizing (180 to 150 μm or 80/100 mesh), and fine grinding (150 to 38 μm or 100/325 mesh). In contrast, sediment samples commonly require only fine grinding. The required final size is usually between 149 and 44 μm (100/325 mesh). Grinding of geological samples and sediments is a batch operation. In the process of reducing bulk geological and sediment samples to powder, frequent sieving of the ground material during grinding removes the finer fractions, speeds up the grinding process, and also ensures that the bulk of the powder will be of the desired size. No material is discarded.

In general, hard materials (e.g., quartz), coarse particles, and fast motion cause wear or abrasion in mills. This wear can cause contamination of the samples, particularly with trace metals. To obtain true concentrations, the sediment samples are split. One split is ground in an agate or ceramic dish, and the other in a metal dish in a disk mill. Both splits are then analyzed for trace elements of interest, the results compared, and the lower values selected as true concentrations. However, if samples are to be used for organic analyses, the metals present in the chamber and disks do not pose any contamination problem.

The choice of grinding equipment generally depends on the quantity of sediment to be ground, hardness of the particular mineral particles, and contamination considerations. Since the sediments are collected wet and if the presence of water is not objectionable, wet grinding can be applied with advantage. In fine dry pulverizing or disintegration steps, surface forces come into action to cause flocculation. An intermediate product in size between 425 and 150 μm (40/100 mesh) is made by removing fine material and screening the coarse material. Grinding of any sample generates grains and particles of different sizes. Moreover, under the same conditions, hard minerals (e.g., quartz, garnets, amphiboles) disintegrate less than soft minerals and rocks (e.g., talc, clays, limestone).

Alternate grinding and sieving is an efficient method to obtain particle uniformity. For example, in the treatment of freeze-dried Great Lakes sediments for archive storage, grinding and sieving were alternated until more than 90% of the sample passed through a sieve opening of 250 μm (60 mesh) (Bourbonniere et al., 1986). However, for analyses requiring grains to be as small as possible (e.g., X-ray fluorescence analyses), the stored sediment will be subjected to another, finer grinding. Sediment reference materials from the National Research Council of Canada (NRCC, 1988) are sieved and/or ground to pass through a 125-μm (120-mesh) sieve.

Mortars and Pestles

A mortar and pestle is an important and indispensable tool in the preparation of sediment samples for testing and analyses. Mortars and pestles of suitable size are commonly used manually, or operated mechanically, to grind small samples to the desired particle size. Manual grinding is time consuming and, consequently, relatively costly. If a homogeneous sample is required, alternate grinding and sieving using small sieves about 50 mm in diameter is essential. The abrasion of the mortars and pestles is greater in mechanical grinding than in manual grinding. Users should take the abrasion into account and assess the potential for contamination from the equipment, particularly with respect to the accuracy of elemental analyses.

Mortars and pestles, in sizes from 35 to 200 mm, are made of various materials: agate, alumina, steel, glass, and porcelain. Agate is a waxy variety of cryptocrystalline quartz with submicroscopic pores. Aluminum oxide mortars and pestles are more resistant to abrasion and less porous than agate. Diamonite mortars are made of finely powdered synthetic sapphire molded under high pressure and sintered at high temperature. Mineral sapphire is blue corundum, nearly pure aluminium oxide with minor amounts of iron and titanium. Porcelain and glass mortars and pestles are considered unsuitable for grinding sediments to be analyzed for trace elements because of contamination. Porcelain mortars and pestles, as well as rubber-tipped pestle and mortars, can be used for breaking up aggregates formed by air- or oven-drying sediments containing clays prior to sieving or mechanical grinding. Alternating grinding and sieving is usually effective for adequate desegregation of the sample and thorough homogenization.

Grinders

There are several commercially available grinders able to reduce sediment samples to 100 μm (150 mesh) and smaller grains. Ball and disc mills are very effective for disintegration, but there is less control over the final particle size. The disk mill is a high-speed disintegration apparatus that breaks agglomerates of various minerals and produces a blend of particles in sizes between 212 and 75 μm (70/200 mesh). For example, a container lined with tungsten carbide or agate, a ring, a solid grinding stone made of the same material, and a cover are used in the swing mill produced by Siebtechnik, Germany (Figure 6-9). Shaterbox, a vibratory disk mill such as that produced by Spex Industries, Michigan, United States, is a similar machine, commonly used in geological and environmental laboratories for fine grinding various materials, including sediments.

Hammer-type mills are designed for capacities from 2 kg/h to several thousand kg/h. Small hammer type mills are suitable for reducting small quantities of fragments <6 mm in size to various degrees of fineness. The mill features a shaft carrying swing hammers, pivoting on a disc that rotates at high speed. Gravity and suction feed material to the mill through a spout in the cover. The pulverized particles are forced through a grate into a collection receptacle. The required grain size is achieved by changing the screens.

Ball and pebble mills have a steel or stone-lined cylindrical steel shell, rotating on a horizontal axis. They contain a charge of steel balls or stone pebbles and the sediment to be reduced. Size reduction is effected by the tumbling of the balls or pebbles on the material between them. The size reduction can be obtained from 3.5 mm (10 mesh) to 45 μm (325 mesh). The

Figure 6-9 Grinder with a grinding dish.

laboratory ball mill is a mechanically rotated steel or ceramic container filled with ceramic balls and sediment sample to about one-third of its volume. It produces a fine powder and mixes thoroughly, but cleaning is inconvenient.

Large quantities of sediment (up to 300 l wet sediment) collected for large-scale tests are usually air-dried and ground using special grinders, such as the Kelly Duplex grinder (Duplex Mill and Manufacturing Company, Springfield, Ohio). This equipment was used for grinding air-dried sediments to pass a 2-mm screen intended for plant bioassays (Folsom et al., 1981).

6.4.4 Mixing and Homogenization

The final goal in processing sediment samples is to produce homogeneous samples that will yield precise results in replicate determinations of inorganic or organic components. Aside from the representativeness of the method used for the original sample collection, the degree of success attained in sample homogenization and splitting is largely responsible for the variability in analytical results. Grant and Pelton (1973) treat the subject of homogeneity and sampling of solids in a theoretical and statistical manner, and the reader is referred to that paper for methods of predicting homogeneity from theoretical considerations. The discussion here will cover the practical aspects and illustrate the success of homogenization with examples.

When one considers the causes of inhomogeneity, the unit that determines segregation becomes very important. The goal in creating a homogeneous sample is to make this unit as small as possible. As an example, a soil or sediment sample that is poorly sorted may contain relatively large sand grains made up of dense minerals mixed with fine clay particles. If this was a dry sample and it was mixed without any treatment, the finer grains would tend to collect in the lower portions of the mixing container. If this was a wet sample, the opposite could occur; the heavier and larger grains would tend to settle first, leaving the finer particles enriched in the upper portions of the container. The

Figure 6-10 Sediment mixer.

unit of segregation in this example is the large size and greater density of the sand grains. Treatment that would be appropriate in this case might be to even off the grain size distribution of the sample by grinding.

Mixing of untreated or ground sediment is required for homogenization and can be carried out by a variety of simple operations:

- Conning and quartering.
- Turning the sample over and over with a spatula.
- Rolling the sediment sample spread on a sheet of paper, plastic, foil, or cloth (see example below).
- Using mechanical rotating mixers consisting of boxes of various shapes, e.g., V-shaped, cones, rectangles, cylinders or cubes on a diagonal (see Figure 6-10).
- Using a Jones riffle splitter for mixing and splitting finely ground material. The sample is poured through the splitter and divided in half. Halves are alternately resplit until the desired homogeneity is obtained.

The following method is modified from one used by the Geological Survey of Canada for preparing geological samples for analyses. It is also useful for mixing sediment samples weighing from 50 g to several kilograms:

- Place the sediment sample on a sheet of glazed paper, plastic, or rubber mat.
- Roll the material from one corner of the sheet to the opposite corner by raising one corner and causing the material to tumble over upon itself; repeat the process by raising each corner in succession until thorough mixing of the sample has been achieved (usually between five and ten times for each corner).
- If the original quantity of the sediment sample needs to be reduced to a particular quantity, such as 5 to 20 g, use the conning and quartering technique. Place the material in a conical shape in the center of the sheet, spread it out into a circular cake, divide it into four quarters, and remove

opposite quarters; repeat mixing, conning, and quartering and remove opposite quarters. A quarter tool can be used for separating the cone into four quarters. Continue until the desired quantity is attained.

- Transfer the final sample to a labelled plastic or glass vial, seal it with a tight cap, and store it until needed.

6.4.5 Procedure for Collection and Preparation of Cores for Sediment Dating

Rates of sediment deposition during the past 100 years is usually determined by different radiometric techniques based on measurements of profiles of either an artificially produced radionuclide, such as ^{137}Cs, or naturally occurring radionuclide, such as ^{210}Pb (Krishnaswamy et al., 1971; Koide et al., 1972; Robbins and Edgington, 1975; Farmer, 1978; Durham and Joshi, 1980; Joshi, 1985; Turner and Delorme, 1988). Sediment dating by ^{137}Cs and ^{210}Pb has been used in assessing the extent of impacts on lakes from human activities, as the input of contaminants into the sediment started in the last century (Durham and Oliver, 1983; Mudroch et al., 1989). Relative time markers in sediment cores, such as pollen grains, have also been used in recent sediment dating. However, these are typically site-specific for a water shed or a larger area (McAndrews, 1976). Horizon markers for determining the age of sediment deposited between the past 300 and 40,000 years include fossil invertebrates, paleobotanical remains, and ^{14}C (Haworth and Lund, 1984).

Collection of sediment cores for determining sediment age and sedimentation rates needs to meet several conditions.

- The selection of a proper location in freshwater and marine environments for collecting cores is critical. Only locations with undisturbed, fine-grained sediment accumulation are suitable.
- The core should be of a sufficient diameter to yield an adequate amount of sediment for analyses.
- The corer should gently penetrate the sediment-water interface without any disturbance of the interface by a shock wave preceding the corer, and without any loss of fine-grained particles. Sediment cores with disturbed top sediment are not suitable for dating.
- The sample interval, typically 1-cm thick, is selected for core subsampling. The top sections of the sample, generally up to 3 cm, consist usually of soft, fine-grained, unconsolidated material with a high water content (up to 95%), and must be subsampled carefully, for example, using a large syringe or a pipette with the wide bottom opening. The remainder of the sediment core can be subsampled into 1 cm sections by the method described in Section 5.6.2. Alternatively, each section around the circumference can be trimmed to minimize the risk of contamination by sediment particles that may have been carried from the surface down the inside of the core tube during penetration into the sediment. However, the quantity of material adhering to the walls of the core tube is so small that this contamination was found negligible (J.A. Robbins, personal communication).

- Dry and wet weights of each sediment section have to be recorded for determining water content.
- The ratio of water content to dry weight of the deepest section of the sediment core is typically the reference value to which all other ratios are normalized.
- The effect of sediment compaction must be allowed for (Robbins and Edgington, 1975). Normalization procedure converting the measured length of an uncompacted sediment section to a theoretical compacted length was described by Farmer (1978).

Procedures for determining sediment age can be carried out only by specialized laboratories, usually at universities or government research facilities. Consequently, these laboratories will issue detailed instructions to the sampling personnel about the handling and preparation of the sediment samples.

The procedure prior to and after sediment collection and preparation for dating by a radiometric method (^{137}Cs or ^{210}Pb) usually involves the following steps:

- Preparation of an appropriate number of labelled, plastic vials or bottles with lids. In measuring radionuclide concentrations, we recommend contacting the personnel of the laboratory equipped with detectors to be used and following their instructions on the size of the vials or bottles that may be placed directly in the detector chamber, saving the time and risk of cross-contamination in transferring the sediment samples from the sampling vials into other containers.
- Weighing the vials without lids and recording their weights.
- Selecting the sampling interval of the recovered sediment core.
- Extruding and subsampling the sediment core into the selected intervals.
- Placing each sediment interval into a preweighed, numbered vial or bottle, and tightly applying the lid to prevent moisture loss.
- In the laboratory, removing the lids and weighing the vials with wet sediment intervals, and recording the weights.
- Drying the sediment after removing the lids in an oven- or freeze-dryer to a constant weight.
- Weighing the vials with dry sediment intervals without lids, and recording the dry weights.
- Preparing a sediment data spreadsheet file.
- Applying the selected method for determining sediment age and sedimentation rates.

6.4.6 Stratigraphic Analyses of Unconsolidated Sediment Cores

Sediment cores can be used for stratigraphic studies. Geological studies of sediment cores recovered from ocean basins include early diagenetic processes, the general nature of climatic fluctuations, the correlation of the marine record to the classical glacial sequence on land, etc. (Moore and Heath, 1978).

Bouma (1969) described in detail many different techniques used in studies of sedimentary structures. Stratigraphic analyses of sediment cores are a useful tool in studies of historical changes in sedimentary processes and sediment geochemistry, and in interpreting results obtained by studies of the input of contaminants into the aquatic environment. Procedures used in such stratigraphic analyses of unconsolidated sediment cores usually include X-radiography of the cores, extruding or splitting the core, and photographing and logging the core for color, texture, and structure. X-radiography of the cores is a nondestructive technique that reveals internal structures and particles that may not be visible to the naked eye, and provides a permanent record of these internal structures of the cores. Details of the principles and use of radiography in sediment studies were described by Krinitzski (1970). The equipment for X-radiography is commercially sold by companies dealing with X-ray instruments.

For visual inspection and different tests, the sediment in a plastic core liner can be split lengthwise by a core liner cutting device similar to that described by Mallik (1986). Also, the core can be extruded on a plastic sheet using a core extruder selected with respect to consistency of the material and length of the core. The surface of the extruded or split core should be cleaned by gently scraping the oxidized surface material (usually a different color than that of the inside of the core) with a wet spatula. Scraping should be done across the core, not lengthwise, to prevent smearing the sediment over the entire core. When taking a photograph of the core, identification labels should be placed showing the top and the bottom of the core, core number, sampling station, time of collection, and a scale (e.g., a ruler placed along the core). For example, a 40- to 45-cm section of the core can be photographed at one time at a distance of 80 cm. A label with the number of the section of the core should be placed near the core when the core is divided into more sections for photographing.

A visual description of the core should contain the following information:

- Length of retained core.
- Equipment used for core collection.
- Name of the operator who collected, handled, and split (or extruded) the core.
- Description of splitting (or extruding) of the core.
- Thickness of the sediment units in the core, which may be based on changes in color using, for example, a Munsell color chart.
- Consistency, for example, described as soupy, soft, medium-firm, firm, stiff, loose, packed, etc.
- Texture (estimated particle size, for example, gravel, sand, silt, clay — or the principal component preceded by the modifier, such as silty clay, sandy clay, etc.).
- Structure (graded bedding, cross bedding, laminates, lenses, varves, etc.), recorded in centimeters.
- Presence of organic matter, shells, and coarse fragments with a description of their type and size.

- Sediment odor, for example, odorless (clean material), or chemical odor (chlorine, petroleum, sulfurous), or decaying organic odor (manure, sewage).
- Appearance of oil, coal dust, ash, etc.
- Presence of carbonates (tested by a drop of 10% hydrochloric acid, which generates effervescence in the presence of carbonates).

Figure 6-11 shows an example of lithologic symbols for core logging (Duncan, 1982) and Figure 6-12 gives examples of core logging sheets. The advantages of logging include the standardization of the description of sediment stratigraphy by different workers and easier interpretation of results.

6.5 PRESERVATION OF SEDIMENT AND PORE WATER SAMPLES

The preservation methods discussed here are by no means exhaustive and complete. Clearly it is impossible to prescribe absolute rules for preventing all possible changes. The objective of the following discussion is to illustrate some of the most common chemical and biological changes that inevitably continue after sample collection. Additional advice should be found in analytical chemical manuals (Standard Methods, 1992; Environment Canada, 1993, 1994).

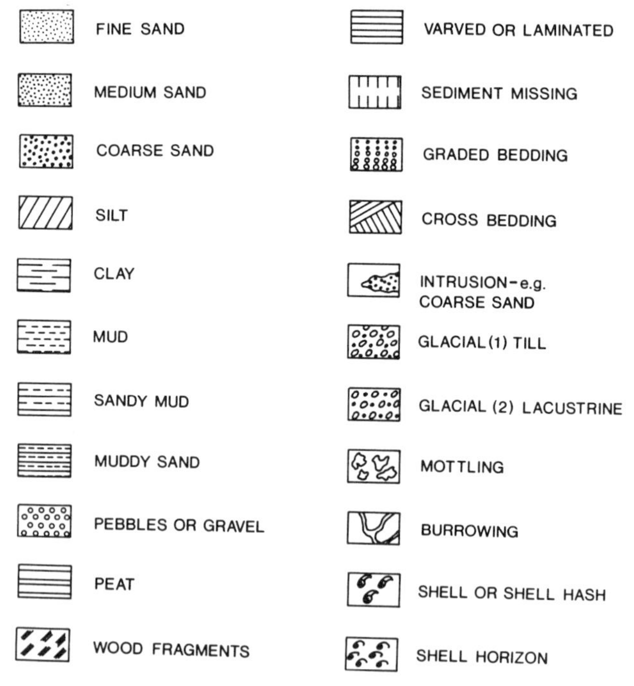

Figure 6-11 Lithologic symbols used in core logging (Duncan 1982).

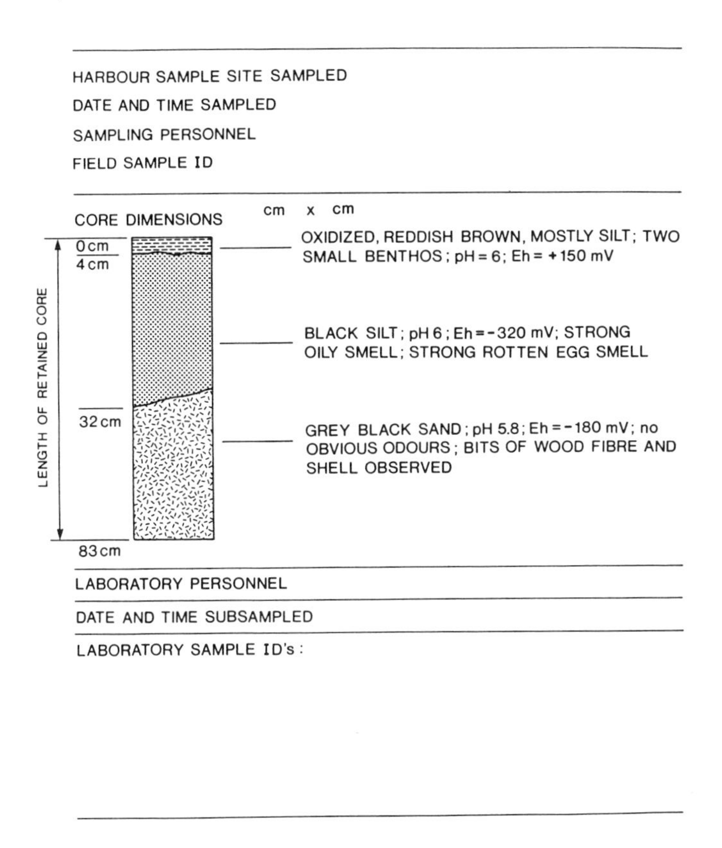

Figure 6-12 Core logging sheet.

Preservation techniques are usually intended to retard microbial degradation, oxidation, and/or loss of volatile components. Methods are limited to pH control, poisoning, drying, refrigeration, freezing, and isolation from the atmosphere. No single preservation method is applicable to all constituents, so it is often necessary to preserve replicate samples or subsamples by different methods when a variety of parameters are required. In general, sample containers should be tightly sealed and headspace should be minimized as soon as the samples are taken. If acid digestion or chemical sequential extractions are required, these procedures should be carried out as soon as possible. Selection of the most appropriate methods should be based on the purpose of the study and the components to be determined (Plumb, 1981). The requirement for multiple samples or splits of single samples increases handling and the time between sampling and preservation; both of these should be kept to a minimum.

Table 6-1 summarizes the preservation requirements for analyses of different parameters in sediment samples. The most common changes in sediments and pore water samples are oxidation, reduction, volatile losses, and biodegradation. Two parameters are important to preserve the integrity of samples: temperature and storage time. Temperature is an important factor that can variously affect the collected sediment samples from the time of sample recovery through handling and processing to the final analyses. Sediment samples intended for analyses or experiments after air- or oven-drying can be stored in containers, cans, plastic bags, etc., at ambient or room temperature. However, sediments collected for determining organic contaminants and mercury should be stored in a refrigerator (about 4°C). The higher the temperature, the higher the risk of losses or changes of volatile compounds. Preservation at low temperatures reduces biodegradation and sometimes volatile loss, but freezing water containing samples can cause degassing, fracture the sample, or cause a slightly immiscible phase to separate (Keith, 1991). The timing of collection, shipment to the laboratory, and analytical work is very important and should be discussed thoroughly during project planning.

Table 6-4 shows the preservation requirements for the analysis of different elements in sediment pore water samples. Immediately after retrieval, pore water samples should be pretreated with the appropriate type and quantity of preservative based on the analysis to be performed. It is recommended to use preservatives only when they are shown not to interfere with the analysis being made. When used, the preservative should be added to the sample container initially so that all sample portions are preserved as soon as collected. It has been shown that temperature and pressure exert a strong influence on the composition of pore water (see Chapter 4). Temperature changes quickly; pH may change significantly in a matter of minutes; dissolved gases (oxygen, carbon dioxide) may be lost. All of these parameters should be determined in the field. With changes in the pH-alkalinity-carbon dioxide balance, calcium carbonate may precipitate and cause a decrease in values for calcium and for total hardness. To avoid loss of volatile species, samples should be collected in a completely filled container, overfilling it before capping or sealing. For nonvolatile species, pore water samples should be collected in preacidified vials to a final pH of 2.5–3.5, usually ultra-pure HNO_3 about 10 μl per 5 ml. The samples should be stored at 4°C, and analyzed within two weeks. Pore water samples collected for determining dissolved organic carbon and methane should be preserved with $HgCl_2$ and analyzed within 3 days.

Biological changes taking place in a sample may change the oxidation state of some constituents. Iron and manganese are readily soluble in their lower oxidation states but relatively insoluble in their higher oxidation states; therefore, they may dissolve from the sediment or precipitate, depending on the redox potential of the sample. Microbial activity may be responsible for changes in the nitrate-nitrite-ammonia content, for decreases in phenol concentration and in biochemical oxygen demand (BOD), or for reducing sulfate to sulfide (Standard Methods, 1992). To minimize the potential for volatilization or biodegradation, samples should be kept as cool as possible without freezing until analysis. Preferably, pack samples in crushed ice before shipment; avoid dry ice because it will freeze samples. If immediate analysis is not possible, storage at 4°C is generally recommended for most samples (Tables 6-1 and 6-4). As a general rule, the shorter the time lapse between collection of a sample and its analysis, the more reliable will be the analytical results. Determination of certain physico-chemical parameters, such as pH, Eh, temperature, and cation exchange capacity, should be made in the field immediately after sample collection.

6.6 STORAGE OF SEDIMENT SAMPLES

The storage conditions of sediments depend on practical aspects and limits as well as previous knowledge of the stability of known or expected contaminants in the sediments. Preservation and storage are two aspects of sample handling that go hand-in-hand. Storage times for different analyses of sediments and sediment pore water are given in Tables 6-1 and 6-4, respectively.

Shipment of collected samples has to be planned, specified, and defined prior to any sediment sampling. It should be general procedure to ship the containers with the samples in coolers filled with ice cubes as soon as possible for processing, preservation, and analyses. Coolers are available in various sizes with volumes ranging from 5 to 60 l, and can be obtained from sources

Table 6-4 Preservation Requirements for Analyzing Elements in Sediment Pore Water Samples

Element	pH of treated sample	Preservative[a]	Sample containers[b]	Time for analysis
Trace metals	2.5–3.5	10 μl of concentrated HNO_3, stored at 4°C	P, T	Within 14 days
Mercury	2.5–3.5	50 μl H_2SO_4, stored at 4°C	G, T	Within 14 days
Nutrients	2.5–3.5	10 μl of 7% H_2SO_4, refrigerated	P, T	ASAP
DOC and methane	Untreated	50 μl of saturated $HgCl_2$	G (Vacutainers)	Within 3 days
Reactive SiO_2	1.5–2.5	10 μl of concentrated HCl	P, T	Within 14 days
pH, Eh, CEC	Untreated	None	P, G, T	Immediately in the field
Phenols	<2	15 μl of 7% H_2SO_4	P, G, T	Within 1 month
Cyanide	<12	NaOH, refrigerated in dark	P, G	Within 24 hours
Organo-phosphates	4.4	Concentrated HCl stored at 4°C	G	Within 1 month
Gases	Untreated	Equilibration to atmospheric pressure with N_2	G (Vacutainers)	Within 3 days

[a] Volumes of preservatives are based in 5-mL samples.

[b] P = polyethylene or polypropylene; G = glass; T = Teflon®.

such as scientific supply companies, hardware or department stores, etc. The personnel in charge of the sediment sampling project must decide in advance about the numbers and sizes of the required coolers with respect to the number of samples to be collected and the numbers and sizes of the containers to be used in the field, as well as the availability of ice cubes, blocks, or dry ice. Storage of samples in coolers is useful for insulation during sampling in winter when freezing the samples has to be avoided.

Because of operational considerations, it may not be possible to store samples in the field in the same way that they will be ultimately stored. A temporary storage method may have to be adopted. One should strive to compromise the samples as little as possible and implement the best storage conditions as soon as possible. For instance, Bourbonniere et al. (1986) stored samples on board ship frozen in solvent-cleaned glass trays at –20°C immediately after sampling from a box core and maintained them at that temperature until freeze-dried. In this example, the temporary storage method (freezing) was also a necessary prerequisite for the ultimate storage method (freeze-drying). In cases where samples must be frozen, dry ice is a good choice. When the receiving laboratory is not ready to process the samples, they must be stored in refrigerators or freezers. In such cases, the holding time is exceeded with unknown consequences.

The American Society for Testing and Materials (1987) defines the holding time as "the period of time during which a water sample can be stored after collection and preservation without significantly affecting the accuracy of analyses." Lacking any guidelines and scientific data dealing with the integrity of sediment samples between collection and analyses or other testing, this definition of the holding time for water samples may be applied to define the holding time for sediment samples.

Samples collected for bioassays should be refrigerated and transported or stored in coolers filled with ice. In cases where these samples are also to be used for chemical analyses, they should be collected and stored in the appropriate containers, as discussed previously in this chapter. While freezing is considered appropriate for sediment samples collected for chemical analyses, it is not recommended for toxicity tests (Malueg et al., 1986), because it can affect the toxicity of sediments.

Biological tests should be conducted as soon as possible, and a 2- to 7-day time frame has been recommended (Swartz et al., 1985; Anderson et al., 1987; Burton and Stemmer, 1987; Klump et al., 1987; Breteler et al., 1989). However, Tatem (1988) showed that some contaminated sediments can be stored at 4°C for up to 12 months without significant changes in toxicity. If, for any reason, sediments collected for bioassays have to be stored for a long time, an easy, short-duration toxicity assay should be carried out periodically to determine to what degree change has occurred. In some studies, sterilization was used to inhibit biological activity in collected sediments. This was done by autoclaving (Clark et al., 1987; Jafvert and Wolfe, 1987), adding antibiotics (Danso et al., 1973; Burton and Stemmer, 1987) or adding chemical inhibitors such as formalin or sodium azide (Wolfe et al., 1986; Jafvert and Wolfe, 1987).

If the vessel used for sediment collection has a freezer and/or refrigerator, the sediment samples are placed in appropriate containers that are immediately transferred to the freezer or refrigerator. The transportation of samples from the freezers and refrigerators on the vessel to a permanent storage room or laboratory is usually in coolers filled with ice. After transporting the samples from the field, storage in refrigerators (usually at 4°C) and/or freezers (usually at –20°C) is essential to preserve the integrity of the collected sediments. Refrigerators and freezers are available from various sources, such as scientific supply companies and department stores. General-purpose laboratory or kitchen upright freezers designed for the storage of pharmaceuticals and chemicals as well as food — which requires a freezing temperature in the range –12 to –30°C — equipped with adjustable epoxy-coated steel shelves, are available in various sizes with volumes from 150 to 750 l. General-purpose upright refrigerators with a temperature range 0 to 14°C, usually set at 4°C by the manufacturer, are available in various sizes with firm or adjustable shelves. For a small laboratory, a freezer/refrigerator can be suitable storage equipment for two sets of sediment samples requiring different storage temperatures, namely 4°C and –20°C. Laboratories involved in a large number of analyses of sediments and/or other materials may have an enclosed, separate room for storing samples at room temperature, and walk-in refrigerators and freezers, such as the walk-in refrigerator in Figure 6-13.

Typically, after being analyzed or tested, the samples in the original, clean containers are either immediately discarded or are kept till the analyst or the customer is satisfied with the results. Simultaneously, the laboratory decides the fate of the containers, either to clean them for further use or, if they are heavily contaminated or disposable, discard them.

A special case is the permanent storage of sediment samples in sediment archives from which they are issued only when required for specific projects. A number of environmental specimen banks were established to store frozen materials at –40°C, –80°C, and –196°C, namely the U.S. National Environ-

Figure 6-13 Sediment cores stored in walk-in refrigerator.

mental Specimen Bank (Wise and Zeisler, 1984; Wise et al., 1984), the environmental specimen bank program of Germany (Kayser et al., 1982; Stoeppler et al.,1984), the Canadian Wildlife Service National Specimen Bank (Elliot, 1984, 1985), and the Great Lakes Biological Tissue Archive (Hyatt et al., 1986). The Great Lakes Sediment Bank, stored at the National Water Research Institute, Environment Canada, Burlington, Ontario (Bourbonniere et al., 1986), contains sediment samples collected with a box corer from a vessel, placed in solvent-cleaned glass trays, frozen on board ship at –20°C, and freeze-dried in the laboratory. Drying times were kept as short as possible. Dry samples were stored at room temperature.

REFERENCES

Ackermann, F., A procedure for correcting the grain size effects in heavy metal analyses of estuarine and coastal sediments, *Environ. Technol. Lett.*, 1, 518, 1980.

Ackermann, F., Bergmann, H., and Schleichert, U., Monitoring of heavy metals in coastal and estuarine sediments — a question of grain size: <20 μm versus <60 μm, *Environ. Technol. Lett.*, 4, 317, 1983.

Adams, D.D., Darby, D.A., and Young, R.J., Selected analytical techniques for characterizing the metal chemistry and geology of fine-grained sediments and interstitial water, in *Contaminants and Sediments*, Vol. 2, Baker, R.A., Ed., Ann Arbor Science Publishers, Ann Arbor, 1980, 3.

American Society for Testing and Materials (ASTM), Annual Book of Standards, *Procedures for Testing Soils*, 4th ed., American Society for Testing and Materials, Philadelphia, 1964.

American Society for Testing and Materials (ASTM), *1969 Book of ASTM Standards*, Pt. 11, American Society for Testing and Materials, Philadelphia, 1969a.

American Society for Testing and Materials (ASTM), *1964 Book of ASTM Standards*, Pts. 9 and 30, American Society for Testing and Materials, Philadelphia, 1969b.

American Society for Testing and Materials (ASTM), *Section 14, General Methods and Instrumentation*, American Society for Testing and Materials, Philadelphia, 1986, 5.

American Society for Testing and Materials (ASTM), *Standard Practice for Estimation of Holding Time for Water Samples Containing Organic Constituents*, American Society for Testing and Materials, ASTM D4515-85, Philadelphia, 1987.

Anderson, J., Birge, W., Gentile, J., Lake, J., Rodgers, J. Jr., and Swartz, R., Biological effects, bioaccumulation, and ecotoxicology of sediment-associated chemicals, in *Fate and Effects of Sediment Bound Chemicals in Aquatic Systems*, Dickson, K., Maki, A., and Brungs, W., Eds., Pergamon Press, New York, 1987, 267.

Birge, W.J., Black, J., Westerman, S., and Francis, P., Toxicity of sediment-associated metals to freshwater organisms: biomonitoring procedures, in *Fate and Effects of Sediment-Bound Chemicals in Aquatic Systems*, Pergamon Press, New York, 1987, 199.

Bouma, A.H., *Methods for the Study of Sedimentary Structures*, John Wiley & Sons, New York, 1969, 458.

Bourbonniere, R.A., VanSickle, B.L., and Mayer, T., *The Great Lakes Sediment Bank — I*, Report 86-151, National Water Research Institute, Environment Canada, Burlington, 1986.

Breteler, R.J., Scott, K.J., and Shepherd, S.P., Application of a new sediment toxicity test using the marine amphipod *Ampelisea abdita* to San Francisco Bay sediments, in *Aquatic Toxicology and Hazard Assessment*, Vol. 12, Cowgill, U.M. and Williams, L.R., Eds., American Society for Testing and Materials, Philadelphia, 1989, 304.

Burton, G.A. and Stemmer, B.L., Factors affecting effluent and sediment toxicity using cladoceran, algae and microbial indicator assays, *Abstract Ann. Meeting Soc. Environ. Toxicol.*, Pensacola, 1987.

Burton, G.A. Jr., Stemmer, B.L., Winks, K.L., Ross, P.E., and Burnett, L.C., A multitrophic level evaluation of sediment toxicity in Wankegan and Indiana Harbours, *Environ. Toxicol. Chem.*, 8, 1057, 1989.

Clark, J.R., Patrick, J.M. Jr., Moore, J.C., and Lores, E.M., Waterborne and sediment-source toxicities of six organic chemicals to grass shrimp (*Palaemonetes pugio*) and amphious (*Branchiostoma caribaeum*), *Arch. Environ. Contam. Toxicol.*, 16, 401, 1987.

Cranston, R.E., Accumulation and distribution of total mercury in estuarine sediments, *Estuarine Coastal Mar. Sci.*, 4, 695, 1976.

Crossland, N.O. and Wolff, C.J.M, Fate and biological effects of pentachlorophenol in outdoor ponds, *Environ. Toxicol. Chem.*, 4, 73, 1985.

Danso, S.K.A., Habte, M., and Alexander, M., Estimating the density of individual bacterial populations introduced into natural ecosystems. *Can. J. Microbiol.*, 19, 1450, 1973.

Day, K.E., National Water Research Institute, personal communication, 1994.

Day, K.E., Kirby, R.S., and Reynoldson, T.B., The effects of sediment manipulations on chronic sediment bioassays with three species of benthic invertebrates, Abstract, *13th Annual Meeting Society of Environment Toxicology and Chemistry*, Cincinnati, Ohio, 1992.

de Groot, A.J. and Zschuppe K.H., Contribution to the standardization of the methods of analysis for heavy metals in sediments, *Rapp. P.-V. Reun. Cons. Int. Explor. Mer.*, 181, 111, 1981.

de Groot, A.J., Zschuppe, K.H., and Salomons, W., Standardization of methods for analysis of heavy metals in sediments, *Hydrobiologia*, 92, 689, 1982.

DeWitt, T.H., Ozrethich, R.J., Swartz, R.C., Lamberson, J.O., Schults, D.W., Ditsworth, G.R., Jones, J.K.P, Hoselton, L., and Smith, L.M., The influence of organic matter quality on the toxicity and partitioning of the sediment-associated fluoranthene, *Environ. Toxicol. Chem.*, 16, 401, 1992.

Ditsworth, G.R., Schults, D.W., and Jones, J.K.P., Preparation of benthic sustrates for sediment toxicity testing, *Environ. Toxicol. Chem.*, 9, 1523, 1990.

Duncan, G.A., *Manual on Procedures for Stratigraphic Analysis of Unconsolidated Sediment Cores*, Report No. 82-24, Hydraulics Division, National Water Research Institute, Environment Canada, Burlington, 1982.

Durham, R.W. and Joshi, S.R., Recent sedimentation rates, Pb-210 fluxes, and particle settling velocities in Lake Huron, *Chem. Geol.*, 31, 53, 1980.

Durham, R.W. and Oliver, B.G., History of Lake Ontario contamination from the Niagara River by sediment radiodating and chlorinated hydrocarbon analysis, *J. Great Lakes Res.*, 9, 160, 1983.

Elliot, J.E., Collecting and archiving wildlife specimens in Canada, in *Environmental Specimen Banking and Monitoring as Related to Banking*, Lewis, R.A., Stein, N. and Lewis, C.W., Eds., Martinus Nijhoff, Boston, 1984, 45.

Elliot, J.E., Specimen banking in support of monitoring for toxic contaminants in Canadian wildlife, in *International Review of Environmental Specimen Banking*, Wise, S.A. and Zeisler, R., Eds., NBS Spec. Publ. 706, U.S. Dept. of Commerce, Washington, 1985, 4.

Environment Canada, *Analytical Methods Manual*, Inland Waters Dir., Water Quality Branch, Ottawa, 1979.

Environment Canada, *Methods Manual for Sediment Characterization*, Saint Lawrence Centre, Quebec, 1993, 145.

Environment Canada, *Summary of Sample Preservation, Container Type and Container Preparation*, The National Labratory of Environmental Testing, Canada Centre for Inland Waters, Burlington, 1994, 19.

Farmer, J.G., The determination of sedimentation rates in Lake Ontario using the Pb-210 method, *Can. J. Earth Sci.*, 15, 431, 1978.

Folk, R.L., *Petrology of Sedimentary Rocks*, Hemphill, Austin, 1974.

Folsom, B.L. Jr., Lee, C.R., and Preston, K.M., Plant Bioassay of Materials from the Blue River Dredging Project, U.S. Army Engineer Waterways Experiment Station, Vicksburg, Miscellaneous Paper EL-81-6, 1981, 24.

Förstner, U. and Salomons, W., Trace metal analysis on polluted sediments. Part I: assessment of sources and intensities, *Environ. Technol. Lett.*, 1, 494, 1980.

Foster, G.D., Baksi, S.M., and Means, J.C., Bioaccumulation of trace organic contaminants from sediment by Baltic clams (*Macoma balthica*) and soft-shell clams (*Mya arenaria*), *Environ. Toxicol. Chem.*, 6, 969, 1987.

Fox, M.E., Murphy, T.P., Thiessen, L.A., and Khan, R.M., Potentially significant underestimation of PAHs in contaminated sediments, in *Proceed. 7th Eastern Region Conference, Quebec*, Canadian Association on Water Pollution Research and Control, 80–86, 1991.

Francis, P.C., Birge, W., and Black, J., Effects of cadmium-enriched sediment on fish and amphibian embryo-larval stages, *Ecotoxicol. Environ. Safety*, 8, 378, 1984.

Gerould, S. and Gloss, S.P., Mayfly-mediated sorption of toxicants into sediments, *Environ. Toxicol. Chem.*, 5, 667, 1986.

Giesy, J.P., Rosiu, C.J., Graney, R.L., and Henry, M.G., Benthic invertebrate bioassays with toxic sediment and pore water, *Environ. Toxicol. Chem.*, 9, 233, 1990.

Giesy, J.P., Graney, R.L., Newsted, J.L., Rosiu, C.J., Benda, A., Kreis, R.G., and Hovarth, F.J., Comparison of three sediment bioassay methods using Detroit River sediments, *Environ. Toxicol. Chem.*, 7, 483, 1988.

Gills, T.E. and Rook, H.L., Specimen bank research at the National Bureau of Standards to insure proper scientific protocols for the sampling, storage and analysis of environmental materials, in *Monitoring Environmental Materials and Specimen Banking*, Luepke, N.P., Ed., Martinus Nijhoff, The Hague, 1979, 263.

Grant, C.L. and Pelton, P.A., Role of homogeneity in powder sampling, in *Sampling, Standards and Homogeneity*, ASTM Spec. Tech. Publ. 540, American Society for Testing and Materials, Philadelphia, 1973, 16.

Harrison, S.H. and LaFleur, P.D., Evaluation of lyophilization for the preconcentration of natural water samples prior to neutron activation analysis, *Anal. Chem.*, 47, 1685, 1975.

Haworth, E.Y. and Lund, J.W.G., *Lake Sediment and Environmental History*, Leicester University Press, 1984, 411.

Hyatt, W.H., Fitzsimons, J.D., Keir, M.J., and Whittle, D.M., *Biological Tissue Archive Studies*, Canadian Tech. Rep. Fish. Aquat. Sci. 1497, Fisheries and Oceans Canada, Burlington, 1986.

Jafvert, C.T. and Wolfe, N.L., Degradation of selected halogenated ethanes in anoxic sediment water systems, *Environ. Toxicol. Chem.*, 6, 827, 1987.

Joshi, S.R., Recent sedimentation rates and Pb-210 fluxes in Georgian Bay and Lake Huron, *Sci. Total Environ.*, 41, 219, 1985.

Kayser, D., Boehringer, U.R., and Schmidt-Bleek, F., The environmental specimen banking project of the Federal Republic of Germany, *Environ. Monit. Assess.*, 1, 241, 1982.

Keilty, T.J., White, D.S., and Landrum, P.F., Sublethal responses to endrin in sediment by *Stylodrilious heringianus* (Lumbriculidae) as measured by a ^{137}Cs marker layer technique, *Aquat. Toxicol.*, 13, 227, 1988.

Keith, L.H., *Environmental Samples and Analyses: A Practical Guide*, Lewis Publishers, Chelsea, Michigan, 1991, 143.

Kersten, M. and Förstner, U., Cadmium associations in freshwater and marine sediment, in *Cadmium in Aquatic Environment*, Nriagu, J.O. and Sprague, J.B., Eds., John Wiley & Sons, New York, 1987, 51.

Klump, J.V., Krezoski, J.R., Smith, M.E., and Kaster, J.L., Dual tracer studies of the assimilation of an organic contaminant from sediments by deposit feeding oligochaetes, *Can. J. Fish. Aquat. Sci.*, 44, 1574, 1987.

Koide, M., Soutar, A., and Goldberg, E.D., Marine geochronology with ^{210}Pb, *Earth Planet. Sci. Lett.*, 14, 442, 1972.

Krinitzski, E.L., *Radiography in the Earth Sciences and Soil Mechanics*, Plenum Press, New York, 1970, 47.

Krishnaswamy, S., Lal, D., Martin, J.M., and Meybeck, M., Geochronology of lake sediments, *Earth Planet. Sci. Lett.*, 11, 407, 1971.

LaFleur, P.D., Retention of mercury when freeze-drying biological materials, *Anal. Chem.*, 45, 1534, 1973.

Landrum, P.F., Eadie, B.J., and Faust, W.R., Variation in the bioavailability of polycyclic aromatic hydrocarbons to the amphipod *Diporeia* (spp.) with sediment aging, *Environ. Toxicol. Chem.*, 11, 1197, 1992.

Landrum, P.F. and Faust, W.R., Effect of variation in sediment composition on the uptake rate coefficient for selected PCB and PAH congeners by the amphipod, *Diporeia sp.*, in *Aquatic Toxicology and Risk Assessment ASTM STP 1124*, American Society for Testing and Materials, Philadelphia, 1991, 263.

Landrum, P.F. and Poore, R., Toxicokinetics of selected xenobiotics in *Hexagenia limbata, J. Great Lakes Res.*, 14, 427, 1988.

Lay, J.P., Schauerte, W., Klein, W., and Korte, F., Influence of tetrachloroethylene on the biota of aquatic systems: toxicity to phyto- and zooplankton species in compartments of a natural pond, *Arch. Environ. Contam. Toxicol.*, 13, 135, 1984.

Luepke, N.P., Final report on workshop, in *Monitoring Environmental Materials and Specimen Banking*, Luepke, N.P., Ed., Martinus Nijhoff, The Hague, 1979a, 1.

Luepke, N.P., State-of-the-art of biological specimen banking in the Federal Republic of Germany, in *Monitoring Environmental Materials and Specimen Banking*, Luepke, N.P., Ed., Martinus Nijhoff, The Hague, 1979b, 403.

Mallik, T.K., An inexpensive hand-operated device for cutting core liners, *Mar. Geol.*, 70, 307, 1986.

Malueg, K.W., Schuytema, G.S., and Krawczyk, D.F., Effects of sample storage on a copper spiked freshwater sediment, *Environ. Toxicol. Chem.*, 5, 248, 1986.

Maxwell, J.A., *Rock and Mineral Analysis*, Interscience Publishers, New York, 1968, 584.

McAndrews, J.H., Fossil history of man's impact on the Canadian flora: an example from southern Ontario, *Can. Bot. Assoc. Bull.*, Suppl. 9, 1, 1976.

McKeague, J.A., *Manual on Soil Sampling and Methods of Analysis*, 2nd ed., Canadian Soc. of Soil Science, Ottawa, 1978.

Moore, T.C. and Heath, G.R., Sea-floor sampling techniques, in *Chemical Oceanography*, Vol. 7, 2nd ed., Riley, J.P. and Chester, R., Eds., Academic Press, New York, 1978, 75.

Mudroch, A. and Duncan, G.A., Distribution of metals in different size fractions of sediment from the Niagara River, *J. Great Lakes Res.*, 12, 117, 1986.

Mudroch, A. and Painter, S., *Comparison of Growth and Metal Uptake of* Cyperus esculentus, Typha latifolia *and* Phragmites communis *Grown in Contaminated Sediments from Two Great Lakes Harbours*, Contribution No. 87-70, National Water Research Institute, Environment Canada, Burlington, 1987.

Mudroch, A. and MacKnight, S.D., *Handbook of Techniques for Aquatic Sediment Sampling*, Lewis Publishers, Chelsea, 1994.

Mudroch, A., Joshi, S.R., Sutherland, D., Mudroch, P., and Dickson, K.M., Geochemistry of sediments in the Back Bay and Yellowknife Bay of the Great Slave Lake, *Environ. Geol. Water Sci.*, 14, 35, 1989.

National Research Council Canada (NRCC), *Marine Reference Materials and Standards*, Circular, National Research Council of Canada, Halifax, 1988.

O'Neill, E.J., Monti, C.A., Prichard, P.H., Bourguin, A.W., and Ahearn, D.G., Effects of lugworms and seagrass on kepone (Chlordecone) distribution in sediment-water laboratory systems, *Environ. Toxicol. Chem.*, 4, 453, 1985.

Perry, J.H., *Chemical Engineers' Handbook*, McGraw-Hill, New York, 1963.

Pillay, K.K.S., Thomas, C.C. Jr., Sondel, J.A., and Hyche, C.M., Determination of mercury in biological and environmental samples by neutron activation analysis, *Anal. Chem.*, 43, 1419, 1971.

Plumb, R.H. Jr., *Procedure for the Handling and Chemical Analysis of Sediment Water Samples*, Tech. Rep. EPA/CE-81-1, U.S. Environmental Protection Agency and U.S. Army Corps of Engineers, Vicksburg, 1981.

Pritchard, P.H., Monti, C.A., O'Neill, E.J., Connolly, J.P., and Ahearn, D.G., Movement of kepone (Chlodecone) across an undisturbed sediment-water interface in laboratory systems, *Environ. Toxicol. Chem.*, 5, 647, 1986.

Robbins, J.A. and Edgington, D.N., Determination of recent sedimentation rates in Lake Michigan using ^{210}Pb and ^{137}Cs, *Geochim. Cosmochim. Acta*, 39, 286, 1975.

Rutledge, P.A. and Fleeger, J.W., Laboratory studies on core sampling with application to subtidal meiobenthos collection, *Limnol. Oceanogr.*, 33, 274, 1988.

Schuytema, G.S., Nelson, P.O., Malueg, K.W., Nebeker, A.V., Krawczyk, D.F., Ratcliff, A.K., and Gakstatter, J.H., Toxicity of cadmium in water and sediment slurries to *Daphnia magna*, *Environ. Toxicol. Chem.*, 3, 293, 1984.

Standard Methods, American Public Health Association, American Water Works Association, and Water Environment Federation, *Standard Methods for the Examination of Water and Wastewater*, 18th ed., Washington, 1992.

Stemmer, B.L., Burton, Jr., G.A., and Leibfritz-Frederick, S., Effect of sediment test variables on selenium toxicity to *Daphnia magna*, *Environ. Toxicol. Chem.*, 9, 381, 1990.

Stephenson, R.R. and Kane, D.F., Persistence and effects of chemicals in small enclosures in ponds, *Arch. Environ. Contam. Toxicol.*, 13, 313, 1984.

Stoeppler, M., Backhaus, F., Schladot, J.D., and Nurnberg, H.W., Concept and operational experiences of the pilot environmental specimen bank program in the Federal Republic of Germany, in *Environmental Specimen Banking and Monitoring as Related to Banking*, Lewis, R.A., Stein, N., and Lewis, C.W., Eds., Martinus Nijhoff, Boston, 1984, 45.

Stone, M. and Mudroch, A., The effect of particle size, chemistry and mineralogy of river sediments on phosphate adsorption, *Environ. Technol. Lett.*, 10, 501, 1989.

Suedel, B.C., Rodgers, J.H. Jr., and Clifford, P.A., Bioavailability of fluoranthene in freshwater sediment toxicity tests, *Environ. Toxicol. Chem.*, 12, 155, 1993.

Swartz, R.C., DeBen, W.A., Jones, J.K., Lamberson, J.O., and Cole, F.A., Phoxocephalid amphipod bioassay for marine sediment toxicity, in *Aquatic Toxicology and Hazard Assessment*, Seventh Symposium, ASTM STP 854, Cardwell, R.D., Purdy, R., and Bahner, R.C., Eds., American Society for Testing and Materials, Philadelphia, 1985, 284.

Swartz, R.C., Schults, D.W., DeWitt, T.H., Ditsworth, G.R., and Lamberston, J.O., Toxicity of fluotanthene in sediment to marine amphipods: a test of the equilibrium partitioning approach to sediment quality criteria, *Environ. Toxicol. Chem.*, 9, 1071, 1990.

Swindoll, C.M. and Applehaus, F.M., Factors influencing the accumulation of sediment-sorbed hexachlorobiphenyl by midge larvae, *Bull. Environ. Contam. Toxicol.*, 39, 1055, 1987.

Taggart, A.F., *Handbook of Mineral Dressing*, John Wiley & Sons, New York, 1945.

Tatem, H.E., Use of *Daphnia magna* and *Mysidopis almyra* to assess sediment toxicity, in *Water Quality '88*, Seminar Proceedings, Willey, R.G., Ed., U.S. Army Corp of Engineers Common Water Quality, Washington, 1988.

Tsushimoto, G., Matsumura, F., and Sago, R., Fate of 2,3,7,8-tetrachlorodebenzo-p-dioxin (TCDD) in an outdoor pond and in model aquatic ecosystems, *Environ. Toxicol. Chem.*, 1, 61, 1982.

Turner, L.J. and Delorme, L.D., *^{210}Pb Dating of Lacustrine Sediments from Hamilton Harbour (Cores 137, 138, 139, 141, 142, 143),* Technical Note LRB-88-9, National Water Research Institute, Environment Canada, Burlington, 1988.

Umlauf, G. and Bierl, R., Distribution of organic micropollutants in different size fractions of sediments and suspended solid particles of the River Rotmain, *Z. Wasser Abwasser Forsch.*, 20, 203, 1987.

Van Driel, W., Smilde, K.W., and van Luit, B., *Comparison of the Heavy Metal Uptake of* Cyperus esculentus *and of Agronomic Plants Grown on Contaminated Dutch Sediments*, Miscellaneous Paper D-83-1, U.S. Army Engineer Waterways Experiment Station, Vicksburg, 1985, 67.

Wall, G.J., Wilding, L.P., and Smeck, N.E., Physical, chemical, and mineralogical properties of fluvial unconsolidated bottom sediments in northwestern Ohio, *J. Environ. Qual.*, 7, 319, 1978.

Warman International Ltd., *Particle Size Analysis in the Sub-sieve Range*, Cyclosizer Instruction Manual, Warman International Ltd., Sydney, Australia, Bulletin WCS/2, 1981.

Water Quality National Laboratory, *Protocol on Preservation, Container Type, and Bottle Preparation*, Water Quality National Laboratory, Environment Canada, Burlington, 1985.

Wilberg, W.G. and Hunter, J.V., The impact of urbanization on the distribution of heavy metals in bottom sediments of the Saddle River, *Water Resour. Bull.*, 15, 790, 1979.

Wise, S.A. and Zeisler, R., The pilot environmental specimen bank program, *Environ. Sci. Technol.*, 18, 302A, 1984.

Wise, S.A., Fitzpatrick, K.A., Harrison, S.H., and Zeisler, R., Operation of the U.S. pilot national environmental specimen bank program, in *Environmental Specimen Banking and Monitoring as Related to Banking*, Lewis, R.A., Stein, N., and Lewis, C.W., Eds., Martinus Nijhoff, Boston, 1984, 108.

Wolfe, N.L., Kitchens, B.E., Macalady, D.L., and Grundl, T.J., Physical and chemical factors that influence the anaerobic degradation of methyl parathion in sediment systems, *Environ. Toxicol. Chem.*, 5, 1019, 1986.

CHAPTER 7

Quality Control of Sediment Sampling

7.1 INTRODUCTION

Information on the concentrations of contaminants in sediments is often the basis for important environmental decisions. Due to improvements in analytical techniques, inaccuracies in the assessment of sediment quality are usually due to an improper sediment sampling design and sampling procedures, instead of errors in analysis of collected sediments. The degree of certainty in the final result of a sampling program depends on the performance of each sediment sampling step, such as sample design, collection, preservation, transport, storage, etc. Consequently, the data obtained in all the intermediate steps of the sampling program have to be reliable. Errors can occur in all steps and the objective of quality control (QC) is to identify and quantify them. The final goal is to correct and minimize individual errors and their cumulative effects. QC procedures reduce and maintain random and systematic errors within tolerable limits, while quality assurance (QA) is the management system that ensures an effective QC system is in place and working as intended (Keith, 1991). In the laboratory, QC consists of the technical, day-to-day activities — such as the use of reference materials, spikes, blanks, etc. — to control and assess the quality of the measurements. It should be noted that QC in the laboratory applies only to possible errors introduced after the collection of sediment samples and their transport to the laboratory. Therefore, sediment sampling procedures are required to have their own QC.

Recently, much attention has been focused on QA/QC in chemical analyses of sediment samples. However, it is more complex to measure the accuracy of sampling the sediments, which are, in most cases, heterogeneous. The following two techniques can be used for QC in sediment sampling. One technique consists of the collection of more than one sediment sample at selected sampling sites using identical sampling equipment, such as multicorers, as well as using identical field subsampling procedures, handling and storage of the samples, and methods for sediment analyses. The results of the analyses will show variations due to the sampling and subsampling techniques, but the heterogeneity of the sediment at the sampling site will still affect the test. The sediment sampler must be selected to suit the sediment texture at the sampling site. In the other QC technique, the collected sample is divided into a few subsamples and each subsample is treated as an individual sample. The results of chemical analyses of all subsamples indicate the variability due to the sampling and analytical techniques and sediment heterogeneity within a single collected sample.

7.2 QUALITY ASSURANCE IN SEDIMENT SAMPLING

A complete sequence of sediment sampling, such as sample collection, preservation, storage, and transport, has to be evaluated to measure and minimize the sources of systematic and accidental errors. Typically, 10% of the samples should be collected in duplicate as a component of the QA/QC plan. Some of the sampling stages may cause considerable variations that cannot be statistically quantified, such as the choice of adequate sampling devices, methods of sample storage, and sample preservation.

Basic requirements to assure the quality of sediment sampling are detailed planning and consistent use of qualified personnel, as well as adequate and well-maintained equipment. A few control sites should be included in a sampling program for investigation of sediment contamination. They should be selected, after historical data review, in areas where the sediment is expected to be uncontaminated. Data obtained at the control sites are important as background values when plotting the distribution and concentration gradients of contaminants. When planning the study, a sufficient number of samples to meet the statistical objectives should always be included. A certain number of extra samples should be collected to allow for unexpected laboratory losses and problems with sample transportation or preservation. Figure 7-1 outlines, in a schematic way, the phases of sediment sampling that deserve special attention and quality control. Discussion of the last two stages (i.e., laboratory analyses and interpretation of the results), although extremely important, is not within the scope of this book. The task of sediment sampling includes collecting and handling the samples, use of sampling equipment and its cleaning, and the storage, preservation, and transport of the samples. Each of these processes can be a potential source of contamination of collected samples, generating considerable errors in the evaluation of sediment quality. The relevance of the QA/QC plan to sampling is that if samples are not representative of the sediment investigated, independent of how good the QA/QC of the analysis is, the analytical information will be largely insignificant.

Many of the approaches used for QA/QC in sediment sampling are based on information available on soil sampling. There are excellent reviews covering the subject of QA/QC in soil samples (Kratochvil et al., 1984; Peterson and Calvin, 1986; Webster and Oliver, 1990; Crepin and Johnson, 1993). The U.S.

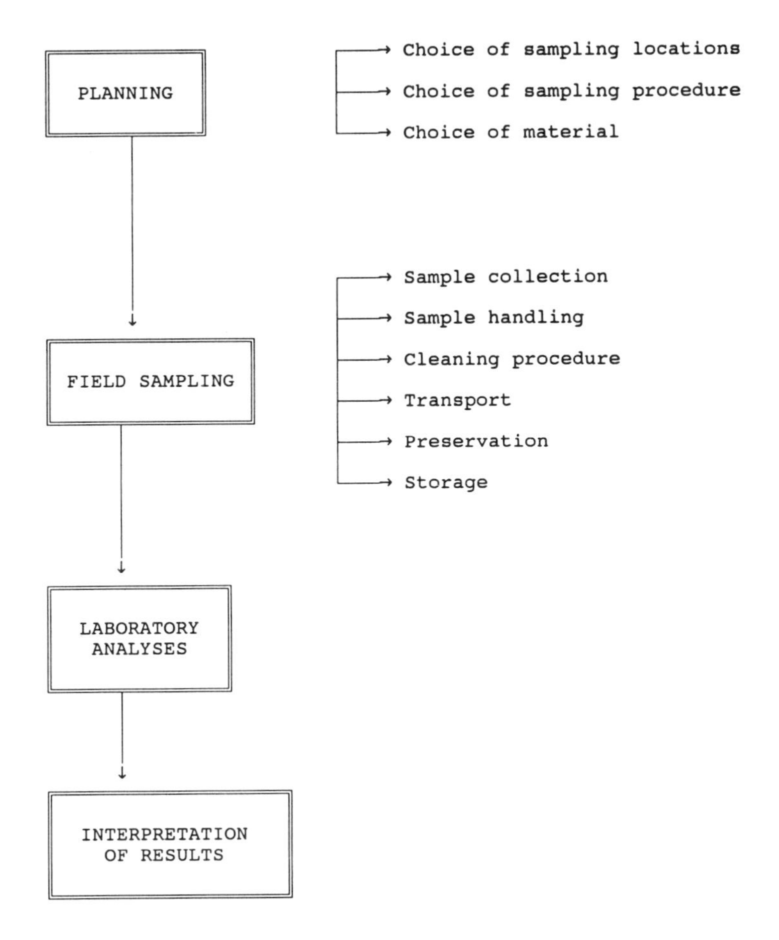

Figure 7-1 Schematic representation of the phases of sediment sampling that deserve special attention and quality control.

Environmental Protection Agency (EPA) has published a manual and software computer program (U.S. EPA, 1991) that provides the foundation for answering the two major questions in QA/QC programs for soil sampling: (1) How many and what type of samples are required to assess the quality of data in a field sampling effort? and, (2) How can information from QA samples be used to identify and control the sources of error and uncertainty in measurement programs?

7.3 PREVENTION OF SAMPLE CONTAMINATION

Sample contamination can be defined as an alteration of the concentration of the parameter of interest resulting from sampling devices in contact with the sample. This alteration can be positive, when the level of the parameter of interest in the sample increases due to its release from the sampler; a negative contamination implies a loss of the parameter of interest from the sample, generally by adsorption onto surfaces or by evaporation. These processes are directly related to the reactivity of the parameter of interest affected by different conditions, such as changes in light intensity and temperature, presence of biological organisms, air, water, interactions with chemicals, etc. The risk of contamination is much greater when sampling sediment pore waters. This is due to considerably lower concentrations (several orders of magnitude) of various parameters in the pore water than in the sediments. However, there are several occasions during sediment sampling where the prevention of contamination deserves special attention:

- When handling of the sample, such as preservation, measuring pH, Eh, etc., must be carried out in the field.
- When the parameters of interest are present at relatively low concentrations, such as cadmium, mercury, and many organic contaminants.
- When contaminated and uncontaminated sediment samples are collected during one field trip, with the potential of generating cross-contamination.
- When the parameter of interest is easily volatilized, such as some forms of mercury and some organic contaminants.

Generally, some simple and inexpensive precautions can considerably improve the quality of the sediment sampling procedure (Laxon and Harrison, 1981; Kosta, 1982; Adeloju and Bond, 1985; Boutron, 1990). When preparing the sampling protocol, it is very important to consider the presence of different elements and compounds in the sampling and laboratory equipment. Sediment sampling involves direct contact of the sample with the sampling device, sample containers, and different reagents used in sample preservation. Sediment samples can be contaminated with pieces of metal paint or corrosive products from samplers or equipment used in handling the samplers. Most samplers are metallic and some may be electroplated or painted to prevent corrosion, particularly when sampling in salt water. Samplers with metal parts painted with cadmium or lead paints are not suitable for collecting sediments for determining metal concentrations. Similarly, the use of oil and grease on the samplers or sampler lifting equipment should be avoided. Sediment samples for quantitative determination of metals or organic contaminants should always be obtained from the center of the collected sample. Plastic liners and core barrels used with gravity corers may be a source of contamination with various organic compounds. However, no data are available on testing contamination of sediment samples collected with liners and core barrels manufactured from different plastic materials. In sediment cores, the material selected for chemical analyses should be collected from the part of the sample that has not been in direct contact with the sampling device, such as the wall of the core tube.

Table 7-1 summarizes some of the data available on the trace metal contents of materials currently used in manufacturing samplers, sample handling equipment, and storage containers. The extent of adsorption/leaching of the elements on/by sample container walls is determined by the chemical composition of the container, geochemical properties of the sample, and length of contact with the sample in the container. The expected levels of the parameters of interest in the sediment samples always needs to be considerably greater than their concentrations in the sampling equipment and containers for sample storage to minimize the risk of contamination. Containers for sediment handling should be carefully selected with consideration of the investigation objectives and the chemical analyses that will be carried out on the sediment samples. Generally, appropriate treatment and handling of laboratory ware made of polyethylene, polypropylene, Teflon®, glass, and silica have to meet the standards required for the determination of traces of contaminants of interest in sediment samples. The selection of containers, preservation, and storage time for different parameters commonly analyzed in sediments is discussed in more detail in Chapter 6.

Table 7-1 Trace Element Content of Materials Commonly Used in Sampling and Storing Sediments and Sediment Pore Water

	Concentration range (µg/g)			
Material	**<0.01**	**0.01–0.1**	**0.1–10**	**10–100**
Polyethylene and polypropylene	Co, Cu, Hg, Sb	Al, As, Cr, Mn Ni, Se, Sr	Cd, Cl, Fe, K, Pb, Si,	Ca, Na, Ti, Zn
Polyvinyl chloride	Se	As, Cu, Co, Sb	Br, Cd, Cr, Mg, Mn, Ni, Pb, Sn, Zn	Al, Ca, Fe, Na, Sn
Teflon®	As, Co, Cs, Pb	Cd, Cr, Co, Cu, Fe, Mn, Ni, Zn	Al, Cl, Na	K, Na
Polycarbonate		As, Cd, Cr, Co, Cu, Mn, Ni, Pb, Zn	Al, Fe	Br, Cl
Borosilicate glass	Ag, Hg, Se, U	As, Au, Co, Se	B, Cd, Cr, Cu, Fe, Ni, Pb, Ti, Zn	Al, K, Mg, Mn, Sr
Silica	Ag, As, Cd, Co Hg, Mn, Mo, Sb, Se	Cr, Cu, Ni	Fe, K, Pb, Zn	
Kimwipe	Ag	Co, Sb	Cr, Fe	Zn
Nucleopore filters	Se	Co, Sb	Cr, Cu, Mn, Zn	Fe
Millipore (HA) filters	Ag, Hg, Se	Co, Sb	Fe, Zn	Cr
Tape	As, Co, Cr, Se	Sb		Fe, Zn

Modified from Murphy (1976); Kosta (1982); Van Loon (1985); Sturgeon and Berman (1987); Azcue (1993).

Handling of sediments in the field, such as some physical and chemical measurements, preservation of subsamples, etc., should be kept to a minimum. Handling may promote potential sources of cross-contamination and alteration of the nature of sediment samples, including deterioration and chemical changes of different compounds and elements. The sampling devices should be cleaned between sampling at different sampling stations to avoid cross-contamination. However, it may be beneficial to keep sediment sampling and subsampling equipment, such as scoops, dishes for homogenization of samples, etc., separate for sampling stations or sediments that are expected to be contaminated from those that are expected to be clean. We recommend washing the sampling equipment by repeated dipping into the ambient water at the sampling site to wash all adhering sediment particles.

7.4 PROCEDURES FOR CLEANING CONTAINERS FOR SAMPLE STORAGE

Problems of contamination of water samples due to either container material or the addition of preservative agents have been widely discussed (Robertson, 1968; Mart, 1982; Ahlers et al., 1990; Nriagu et al., 1992). However, it should be noted that concentrations in water of elements of concern such as arsenic, cadmium, chromium, and lead are in the ng/g range, whereas concentrations of the same elements in the sediments are usually in the μg/g range. Consequently, contamination of water samples by the container's material is by far a more serious problem than is contamination of sediment samples. However, the risk of cross-contamination can be greater when reusing the same container for sample collection. There is no universal cleaning procedure; a specific cleaning method can be very effective for one element, but not sufficient for another. The rigorous cleaning procedures outlined below may not be necessary if the elements of interest in the sediment samples are expected to be present in high concentrations. Thus, the choice of the cleaning procedure, based on the objectives of the study and expected levels of the parameter of interest, must be left to the judgment of the scientist in charge of the sampling program and storage of the sediment samples.

A minimum cleaning procedure should consist of first washing the interior of the containers with hot water and laboratory-grade soap, with subsequent rinsing by deionized water followed either by washing with an acid such as dilute nitric acid for samples selected for determining inorganic compounds, or by rinsing with solvents such as methyl alcohol, dichloromethene, etc., and drying for samples selected for determining organic contaminants. When using containers made of borosilicate glass for samples selected for the determination of organic contaminants, we recommended heating the containers overnight to 550°C. However, specific cleaning procedures should be applied for containers destined for determining inorganic compounds and organic contaminants, as

Table 7-2 Cleaning Procedure for Containers to Hold Sediments Destined for Determination of Inorganic Constituents

1. Scrub with phosphate-free soap and hot water.
2. Wash with high-pressure tap water.
3. Degrease in Versa Clean (Fisher) soap bath for 24 hours.
4. Subject to a 72-hour acid bath with reagent grade 6M nitric acid; drain off acid and rinse with hot water.
5. Rinse with double-distilled water and allow to dry in a particle-free environment.
6. Place containers in heavy polyethylene bags.

Modified from Nriagu et al. (1992).

Table 7-3 Cleaning Procedure for Containers to Hold Sediments Destined for Determination of Organic Constituents

1. Scrub with phosphate-free soap and hot water.
2. Wash with high-pressure tap water.
3. Clean with detergent such as Versa Clean (Fisher) or similar.
4. Rinse three times with organic-free water.
5. Rinse twice with methylalcohol.
6. Rinse twice which dichlomethane.
7. Dry in an oven at 360°C for at least 6 hours.

outlined in Tables 7-2 and 7-3. In contrast to other sources of contamination during the sampling procedures, contributions from reagents can be quantitatively measured. The procedures for cleaning equipment for sampling sediment pore water are described in Chapter 4.

For the long-term storage of sediment samples, plastic containers could be a source of contamination by trace elements due to leaching (Moody and Linstrom, 1977). Properly cleaned containers made of Teflon® and polyethylene are best for storage of sediments collected for determining inorganic constituents (Massee et al., 1981). The containers are first scrubbed with phosphate-free soap and hot water, and then washed with high-pressure tap water and degreased with detergent for 24 hours (Table 7-2). Next, they are subjected to a 72-hour acid bath with 6 M nitric acid. The containers are finally rinsed with hot water and DDW. To minimize contamination, the containers should be kept in plastic bags after the final rinse until sample collection. The treatment summarized in Table 7-2 applies to all new containers as well as those made from material of unknown composition. Once the container undergoes the treatment, steps two and three are not required in subsequent cleanings.

Containers used for sediments to be analyzed for organic components should be carefully chosen; plastic should be avoided wherever possible. Containers (glass, porcelain, stainless steel, or Teflon®) intended for sediments to be analyzed for organic compounds are first washed with high pressure tap water (Table 7-3). The containers are then cleaned with detergent and rinsed three times with water free of organic contaminants. At Environment Canada's National Water Research Institute, in Burlington, Ontario, following water rinsing, containers are rinsed twice with methylalcohol, which should remove residual water and dissolve any polar contaminants. Next, the containers are rinsed twice with dichloromethane to solubilize nonpolar contaminants. Finally, they are dried overnight in a oven at 360°C to completely remove any residues of dichloromethane.

For determining some elements or compounds, specific cleaning procedures are required in addition to those described in Tables 7-2 and 7-3. For example, sample containers for sediment samples collected for biological testing should be sterilized in an autoclave instead of being dried; hot air sterilization at 160°C to 180°C is usually recommended. After sterilization, a thin sheet of sterile paper or similar material should be inserted between the lid and the neck of the container and the top, and covered with aluminum foil or sterilized paper with a cord or rubber band (rubber stoppers cannot be decontaminated by heat treatment).

Special attention must be paid to sediment samples collected in one container and used for the multiple determination of different contaminants. In such a case, the cleaning procedure can be a source of contamination for some of the parameters of interest. For example, contamination problems have been reported in the determination of chromium when sodium dichromate solution was used to clean glass containers, or nitrate contamination introduced by washing the containers with nitric acid, and phosphate contamination introduced by washing the containers with phosphate-containing detergents (Standard Methods, 1992).

7.5 DOCUMENTATION OF SAMPLE COLLECTION

Documentation of collection and analysis of environmental samples requires all the information necessary to (1) trace a sample from the field to the final result of analysis; (2) describe the sampling and analytical methodology; and (3) describe the QA/QC program (Keith et al., 1983).

Correct and complete field notes are absolutely necessary in any sampling program. Poor or incomplete documentation of sample collection can make analytical results impossible to interpret. Figure 7-2 shows a typical form for documenting sediment collection. The following items should be recorded at the time of the sediment sampling:

1. Project or client number.
2. Name of sampling site and sample number.
3. Time and date of sediment collection.
4. Weather conditions, particularly wind strength and direction, air and water temperature, snow or ice cover, thickness of ice when sampling from the ice.
5. Positioning information (equipment used for positioning, any problems encountered during positioning of a station, drawings of sampling site's positions on a chart).
6. Type of vessel used (size, power, engine type).

Project/Client Number: ____________________

Time and Date of Collection: ____________________

Sampling Personnel: ____________________

Weather Conditions: ____________________

Sample identification:

* sampling site: ____________________
* sample number: ____________________
* sediment description: ____________________

Method of collection:

* sampling equipment: ____________________
* vessel used: ____________________
* number and types of containers: ____________________

Notes:

____

____

____

Figure 7-2 Example of a form used in documenting sediment sampling.

7. Type of sediment sampler used (grab, corer) and modifications made on the sampler during sampling.
8. Names of sampling personnel.
9. Notes of unusual events that occurred during sampling. For example, a grab sampler not completely closed, the top of the recovered corer smeared with sediment, the loss of part of a sediment sample from a grab sampler or a corer, the loss of a section of a sediment core through the bottom of the core liner before capping the bottom of the tube, problems with the sampling equipment, or observations of possible sample contamination.
10. Sediment description including texture and consistency, color, odor, estimate of quantity of recovered samples by a grab sampler, length and appearance of recovered sediment cores.
11. Notes on further processing of sediment samples in the field, particularly subsampling methods, type of containers and temperature used for sample storage, and record of any measurements made in the field, such as pH, Eh.

Bound notebooks are preferred to the loose-leaf type and should be kept in a room or in a container that will protect against fire or water damage. Whenever legal or regulatory objectives are involved, notebook data should be entered in ink, each page should be signed and witnessed, and all errors or changes should be struck through one time (to keep them readable) and initialed (Keith, 1991). We also recommend keeping records regarding all personnel involved in sampling, handling, storage, and analysis of the samples.

7.6 FIELD BLANKS AND BACKGROUND SAMPLES

The field QA/QC samples are defined as matrixes containing insignificant or unmeasurable levels of the parameter of interest. Whenever there is a chance of introducing an element or compound that is not a constituent of the sediment sample, it is necessary to ensure control to detect and measure the element or compound. The control samples used in sediment studies include control samples of sampling, transport, sampling equipment, etc., and control samples for laboratory procedures. Requirements for blanks are generally determined by the objectives of the study, taking costs into account. The main objective of the blanks is to detect and quantify the extraneous material introduced into the sample. Sediment sampling generally does not require the use of blanks. However, there are several types of blanks commonly used in sediment pore water sampling:

- Field blanks are samples of laboratory reagents or reference materials that are carried to the field and exposed to the same procedures, such as transfer into containers, field physico-chemical measurements, etc., as the actual samples. They measure incidental or accidental contamination during the entire sampling process.
- Transport blanks are samples free of contamination that are transported from the laboratory to the field and back to the laboratory without being opened. They are used to detect any possible cross-contamination between the containers and the preservative during transport and storage. This kind of blank is more important when the sample is stored for several weeks prior to analysis during which leaching/adsorption of materials from/onto the containers and volatilization of the chemical of interest can occur. One transport blank should be employed per each field trip and each type of sample.
- Equipment blanks are samples of water that have been used to rinse the sampling equipment. These blanks are critical when sampling sediment pore waters, and they are collected before and after cleaning the sediment pore water sampling devices, outlined in Chapter 4.
- Spiked samples in the field are samples to which a known amount of a certain element or compound of interest is added in the field. These samples are used to identify possible interferences of complex matrices, or time-related losses by volatilization.

There are also laboratory blanks, i.e., reagents and methodology. However, discussion of the laboratory blanks is not within the scope of this book.

Background or control samples in environmental studies are samples taken under the same conditions in an area geographically close to the study area, where the parameter of interest is present in concentrations considered natural. These types of controls are critical, due to the great geographic variety of the geological matrixes and heterogeneity of the sediments. Geological areas with anomalous levels of metals and trace elements are common. For example, the average concentration of arsenic in soils is 3 to 5 μg/g, but there are many uncontaminated areas where there are concentrations of arsenic up to 60 μg/g (Nriagu and Azcue, 1990). A similar distribution applies to many other elements.

Comparison of the levels of elements and compounds in the study area to natural levels is fundamental in most environmental studies of sediments. Special attention has to be paid to assure that the control sediment samples came from an area geologically similar to the study area. Further, they have to be collected and analyzed under conditions identical to those used in the collection and analyses of the samples from the study area. Documenting the natural levels of elements and compounds is instrumental to demonstrating the degree of contamination. Therefore, information on the geology at the study area needs to be considered before choosing control sites and interpreting the results. The selection of a proper control site and adequate interpretation of the results are one of the most complex tasks in sediment sampling. In the sampling program, the samples of expected natural concentrations of elements from the control site should be taken prior to those from areas with contaminated sediments to avoid cross-contamination.

Generally, analyzing collected samples is more expensive than analyzing a proper collection of blanks. Consequently, every field investigation should include a full range of blanks, such as transport, equipment, reagents, blanks, etc. However, the field blanks need to be analyzed prior to any of the other blanks, such as transport and equipment blanks, or samples. In case the field blanks do not indicate any type of contamination, the other blanks can be discarded. Only when the field blanks indicate a contamination problem should the other blanks be analyzed to locate the source. These blanks may identify unexpected common sources of pollution, such as contamination of distilled water or reagents, contamination problems associated with transport, and contamination induced during the sampling procedure.

Due to the complex matrix of sediment samples, the determination of representative field blanks is much more complex than in water sampling. As mentioned above, the handling of the samples in the field, such as transferring into containers, sampling, making physico-chemical measurements, storing, etc., presents potential sources of contamination. Therefore, the field blanks have to be treated in the same way as the actual samples, including use of the same sampling devices, containers for sample storage, and preservation techniques.

7.7 PREPARATION OF REFERENCE MATERIALS

It was shown that the sediment type (such as arenaceous, argillaceous, calcareous, sand, silt, clay) and mineral composition play an important role in the distribution of elements in the sediment (Hamilton, 1993). The lack of matrix matching between sediment samples and analytical standards can cause generation of biased data and erroneous conclusions (Massee et al., 1981). Reference materials are stable, homogeneous, and well-characterized materials prepared in quantity, having essentially an identical or similar matrix to the samples collected in a study (Lawrence et al., 1982). To determine the degree of accuracy in a measurement process, the same technique is used to analyze reference materials having similar matrices and concentration levels to the sediment of interest. Reference materials are used to help ensure the accuracy of many applications such as validation of methods, inter- and intra-laboratory quality control, environmental monitoring and surveillance, and environmental research.

Because of the high price of commercially available standard reference materials, it is advantageous for laboratories involved in performing many analyses of sediments to prepare an in-house set of reference materials for different analyses. To prepare sediment reference material, Cheam and Chau (1984) carried out the following steps.

1. *Selection of a Suitable Sampling Site.* Ideally, all certified reference materials should be made from naturally contaminated samples to reflect real environmental situations. The sampling locations were determined by reviewing previous background data to narrow down the sites available to a few sampling stations. Preliminary small samples were then taken to ensure the presence of the parameters of interest in suitable concentrations.
2. *Sample Collection and Identification of Certified Reference Material.* About 500 kg of surface sediments were collected from the chosen site and stored at 4°C in covered galvanized garbage pails precleaned with detergent and organic solvents before use. The sediment was labelled (i.e., WQB-1), and its complete geochemical composition determined.
3. *Preliminary Drying.* This step was important in order to remove about 50% of the water from the sediment. It consisted in freezing the sediment at –20°C for four days. The containers were then taken to room temperature and allowed to thaw; numerous 5-mm diameter holes were drilled on the sides of the pails so that the water could drain out leaving the sediment in a semi-dried state.
4. *Freeze-Drying.* The partially dried sediment was transferred in 20-kg lots to a commercial freeze-drying chamber and dried at reduced pressure and elevated temperature. After drying, the sediment was in the form of small aggregates and weighed 205 kg.

5. *Homogenization.* The aggregate sample sediment was crushed in a Denver roller and then passed through a 125-μm (120-mesh) vibrating screen. The oversized fraction was set aside. The remainder was passed through a 45-μm (325-mesh) vibrating screen and the fraction passing through was collected. The oversize fraction was ground in a ball mill for 1.5 hours and the material was passed through the 45-μm screen again. The oversize fraction was rejected at this stage. The combined ≤45-μm sediment fractions were tumbled in one lot for approximately 8 hours in a 570-l conical steel shell blender. At the end of tumbling, six 50-g samples were removed from the top, middle, and bottom level of the bulk sediment for homogeneity testing.
6. *Homogeneity Test on Bulk Sediment.* Five replicate analyses were made on each sample for each element. Two-way analysis of variance (ANOVA) was performed on the matrices of analytical results (one matrix per element) to test the homogeneity of the material. If the analysis showed the material to be not yet homogeneous, the blending and test for homogeneity were continued until the ANOVA provided evidence of material homogeneity between and within samples, at which stage the material was considered sufficiently homogeneous to be subsampled.
7. *Subsampling and Homogeneity Test on Subsamples.* To avoid settling, the sediment was blended for 1 hour before bottling 25-g sediment samples into 100-ml brown bottles equipped with plastic screw caps. The shiny side of an 8 cm × 8 cm aluminum foil liner, prewashed with petroleum ether or ethyl ether, was inserted under the cap before tightly sealing the bottle to avoid moisture absorption. To ensure further homogeneity of subsamples, one sample was selected out of every 50 samples for determining the parameters of interest. The homogeneity criterion was set at ±5% coefficient of variation by a single-method-single-parameter testing. All bottles were placed in cardboard boxes, sealed with thick plastic bags, and stored at –20°C ready for use.

The success of homogenization in the 75- to 45-μm range of reference materials prepared from sites within the Laurentian Great Lakes is indicated by the data for arsenic, selenium, and mercury in Table 7-4. The authors of these studies have shown that these samples were homogeneous for those elements when tested at 95% significance level according to a two-way ANOVA. A relative standard deviation (RSD) criterion of ±5% of the mean was chosen as a homogeneity criterion for single-parameter-single-method testing.

The prime requirement was that the sample be analyzed many times and, if possible, by different analysts using various methods of determination, such as X-ray fluorescence spectrometry, neutron activation, inductively coupled plasma, and atomic absorption spectrometry for the determination of major and trace elements. The results of these analyses were used to calculate a consensus reference value for each parameter measured. The certification procedure requires:

- at least 120 acceptable determinations for each element;
- a minimum of two independent methods;
- a minimum of three independent laboratories for each element;

Table 7-4 Arsenic, Selenium, and Mercury from Homogenized Bulk Sediment Reference Materials Prepared from Great Lakes Sites

Sample set[a]	*n*	Mean concentration[b]	SD	RSD[c]
Arsenic				
WQB-1	48	22.71	0.53	2.33
WQB-3	30	19.1	0.80	4.19
Selenium				
WQB-1	48	1.07	0.05	4.67
WQB-3	30	1.25	0.04	3.20
Mercury				
WQB-1	30	1.08	0.05	4.63
WQB-1	48	1.08	0.03	2.78
WQB-3	30	2.95	0.11	3.73

[a] WQB-1 sediment data from Tessier et al. (1979); WQB-3 sediment data from Jenne and Louma (1977).

[b] Concentrations in μg/g.

[c] Relative standard deviation expressed as % of mean.

- analysis by each laboratory of two different sets of samples at two different times without prior knowledge of sample type;
- the recommended values for each element to be given at 95% tolerance limit (Cheam et al., 1989).

Certified materials are intensively characterized reference materials. The certified values are obtained by repetitive analysis by several operators using different methods in several qualified laboratories of known precision and accuracy (Rasberry and Gills, 1991). Table 7-5 shows several sediment reference materials presently available commercially. The materials are divided by the certified concentrations of organic and inorganic parameters, with information on the source and concentration ranges. In addition to pollutants, other elements are normally certified in the reference materials. Table 7-6 contains the addresses of some of the organizations that have issued sediment reference and control materials. Additional information regarding standard reference materials can be found in Van Loon and Barefoot (1989).

Participation in domestic and international intercalibration programs has been used by laboratories to gain recognition of their competence in handling and analyzing aquatic sediments. As part of a comprehensive quality control scheme, an in-house prepared standard can then be run with every batch of samples. However, it is also recommended that commercial standard reference materials be used on a weekly basis, at least as an aid to monitoring analytical quality assurance.

7.8 GUIDELINES FOR SEDIMENT QUALITY

A number of international agencies and national jurisdictions have developed sediment quality criteria. However, numerical guidelines defining the

Table 7-5 Examples of Sediment Reference Materials Currently Available (Concentrations in μg/g)

Reference material identification	Sediment source	Parameters[d]	Concentration ranges
Organics			
EC-1[a]	Hamilton Harbour	PAH, PB	5–20
EC-2[a]	Lake Ontario	PAH, PB, CB	0.5–3
EC-3[a]	Niagara River	PAH, PB, CB	0.05–0.9
EC-4[a]	Toronto Harbour	PAH, PB	0.2–1.0
EC-5[a]	Humber River	PAH, PB	0.1–0.5
EC-6[a]	Lake Erie	PAH, PB, CB	0.01–0.2
EC-7[a]	Lake St. Clair	PAH, PB, CB	0.02–0.5
HS-1 to 6[b]	Nova Scotian harbours	PAH, PB	0.14–60
SRM-1939[c]	River sediment	PB	1.07–4.2
SRM-1941[c]	Marine sediment	PAH	0.2–1.2
Inorganics			
WQB-1[a]	Lake Ontario	As, Se, Hg	1–2, 000
WQB-2[a]	Lake Ontario	As, Se, Hg, trace metals	1–2, 000
WQB-3[a]	Hamilton Harbour	As, Se, Hg, trace metals	1–1, 400
SUD-1[a]	Sudbury	Trace metals	0.1–1, 000
TH-1[a]	Toronto Harbour	Trace metals	1–1500
HR-1[a]	Humber River	Trace metals	0.3–1100
MESS-1[b]	Miramichi River estuary	As, Se, Hg, trace metals	0.17–513
BCSS-1[b]	Baie des Chaleurs	As, Se, Hg, trace metals	0.13–229
SRM-1645[c]	Indiana Harbor Canal (Indiana)	As, Hg, trace metals	0.17–714
SRM-1646[c]	Chesapeake Bay	As, Hg, trace metals	0.06–138
SRM-2704[c]	Buffalo River	As, Hg, trace metals	1.2–555

[a] National Water Research Institute (1992).

[b] National Research Council of Canada (1988).

[c] National Bureau of Standards (1992).

[d] PAH = polynuclear aromatic hydrocarbons; PB = polychlorinated biphenols; CB = chlorobenzenes.

levels of chemical contaminants in sediments that are significant from an environmental or human-health standpoint have been established for only a few contaminants. The historical beginning of sediment guidelines was in response to the need to manage dredged materials.

Giesy and Hoke (1990) reviewed the procedures proposed (based on both chemical and biological analyses) as the basis for developing sediment quality criteria. They include

- background concentration approaches;
- water-quality criteria approach;
- equilibrium partitioning approach;
- field bioassay approach;
- screening level concentration approach;
- apparent effects threshold approach;
- spiked bioassay approach.

For the background concentration approach, the concentrations of contaminants in sediments at a particular location are compared with concentrations

Table 7-6 Suppliers of Sediment Reference and Control Materials for Quality Control and Related Environmental Studies

National Research Council of Canada, Division of Chemistry Ottawa, ON K1A OR6 Canada	National Water Research Institute Aquatic Quality Assurance Program Research & Applications Branch Environment Canada, Burlington
Canada Centre for Mineral and Energy Technology 555 Booth Street, Ottawa, Ontario, K1A OG1, Canada	U.S. Geological Survey 12201 Sunrise Valley Dr. Reston, VA 22092, U.S.A.
U.S. Environmental Protection Agency Quality Assurance Branch, Cincinnati, OH 45268, U.S.A.	National Bureau of Standards U.S. Department of Commerce Gathersburg, MD 20899, U.S.A.
National Institute for Environmental Studies Japan Environmental Agency, Tsukuba, Ibaraki 300-21, Japan	Bureau of Analyzed Standards Newham Hall, Middlesborough, Cleveland TS8 9EA, England
Commission of the European Communities Community Bureau of Reference 200 Rue de la Loi, B-1049 Brussels, Belgium	International Atomic Energy Agency Analytical Quality Control Services, P.O. Box 100, A-1400, Vienna, Austria
Polish Academy of Sciences Commission of Trace Analysis of the Committee for Analytical Chemistry - ul. Dorodna 16, Warszawa 030195, Poland	Zentrales Geologisches Institut Invali denstrasse 44 DDR-104 Berlin, Germany
	Association nationale de la recherche technique B.P.20 - 54 501 Vandoenvre-Nancy CEDEX, France

from reference materials (background sites), where contaminant levels are considered to be acceptable based on the presence of indicator organisms (Mudroch et al., 1988). Müller (1979) developed the "index of geoaccumulation" based on background concentrations. In this approach, the concentration of pollutants or nutrients are normalized with respect to grain size, allowing for the comparison of sediment samples taken from different aquatic ecosystems. Förstner et al. (1990) recommended that the sediment properties should be classified on the basis of the carbonate and sulfide inventory, whereas the pollutant load is advantageously assessed by the accumulation rate multiplied with a toxicity factor for the respective substance. However, the sediment background approach has several limitations, such as the potential for inconsistency in choice of suitable reference areas (i.e., definition of background), the variability of natural metal concentrations among geographical regions, the lack of a strong biological basis for the concentrations, and unsuitability of this method for synthetic organic compounds.

For the water-quality criteria approach, concentrations of individual contaminants in sediment pore water are compared to existing water-quality criteria. The main limitation of this approach is that the water quality criteria have been developed for sediment-free systems. Therefore, they do not consider adsorption and bioavailability of the contaminants in the sediments, and may be inappropriate to the sediment pore water environment.

The equilibrium partitioning approach is based on the assumption that the distribution of contaminants among different compartments in the sediment is controlled by a continuous equilibrium exchange among sediment, sediment pore water, and overlying water. This approach suffers from the same limitations as all of the chemically based global and numerical sediment quality criteria in that it does not consider unmeasured chemicals or interactions among chemicals for which there is no toxicological information (Giesy and Hoke, 1990).

The screening level concentration and the apparent effects threshold approaches compare the distribution of benthic invertebrates with the concentrations of contaminants in identical sediments. The benthic invertebrates reside in or on the sediments; they are sessile; their populations are relatively stable in time, with life cycles of one or more years; their taxonomy is reasonably well established; and their responses to environmental changes have been studied extensively (Reynoldson and Zarull, 1993). Specific contaminant concentrations in sediments can be related to the presence or absence of benthic species and their densities and diversity. Criteria established by using these methods would also need to be validated for each contaminant by the use of field bioassay and toxicity testing. The criteria from the field bioassay approach are based on dose-response relationships developed by exposing benthic organisms to collected natural sediments with known concentrations of contaminants and measuring mortality and sublethal effects of the test species. The sediment quality criteria values are established at contaminant concentrations that correspond with a statistically significant difference in mortality or other biological response between a test sediment and a control sediment. This approach considers sediment toxicity as a measurement of the total effect of all toxic agents and is useful for identifying sediments of environmental concern, but requires integration with other approaches to provide contaminant-specific sediment quality criteria values. However, the major objections still posed to employing benthic community structure analysis for setting sediment guidelines is their lack of universality (i.e., they are completely site-specific) and the inability to establish quantitative objectives for their application (i.e., what should the community "look" like?) (Reynoldson and Zarull, 1993). For the spiked bioassay approach, dose-response relationships are determined by exposing test organisms to sediments that have been spiked with known amounts of contaminants (individually or in combination).

There is no unique criteria for selecting a sediment quality guideline method. Recently, initiatives and reviews in the United States and Canada (U.S. EPA, 1984; Beak Consultants, 1987; International Joint Commission, 1987, 1988; Dutka et al., 1988; Shea, 1988; Persuad et al., 1989; Giesy and Hoke, 1990; Reynoldson and Zarull, 1993), in Germany (Förstner et al., 1990), in Sweden (Håkanson, 1980), and in the Netherlands (van de Guchte and van

Leeuwen, 1988) describe current thinking on methods of developing criteria and\or objectives. Giesy and Hoke (1990) concluded that "the chemically based methods may be useful for setting global target guidelines but should be supplemented with biologically based local or regional criteria." Hart et al. (1988) recommended determining potential sediment quality criteria using as many approaches as possible to establish whether there is some consensus of appropriate criteria among the approaches, and then choose an appropriate criterion. Using total concentration pollutants in sediments as the role criterion for determining sediment quality does not necessarily reflect the risks to the environment. Analyzing total concentrations of trace elements in sediment gives an indication of their significance but not their availability to living organisms.

The first sediment quality guidelines were developed by the Federal Water Quality Administration and adopted by the United States Environmental Protection Agency (U.S. EPA) in 1973 (Anon, 1973). This included seven parameters, and if any one of the numerical values was exceeded, the sediment was classified as polluted and had to be treated as such. Subsequently, Region V of the U.S. EPA (1977) developed guidelines for evaluating Great Lakes harbor sediments, based on total contaminant concentrations of sediments. These guidelines, developed to address the disposal of dredged material, have not yet been adequately related to the impact of the sediments to the lakes, and are considered as interim guidelines until more scientifically sound guidelines are developed (Fitchko, 1989). Based on the concentrations of contaminants, the

Table 7-7 Guidelines for Pollution Classification of Great Lakes Harbor Sediments (concentrations in µg/g)

	Nonpolluted (less than)	Moderately polluted	Heavily polluted (greater than)
Volatile solids	5%	5–8%	8%
Chemical oxygen demand	40,000	40,000–80,000	80,000
Oil and grease	1,000	1,000–2,000	2,000
Total Kjeldahl nitrogen	1,000	1,000–2,000	2,000
Ammonia	75	75–200	200
Cyanide	0.1	0.1–0.25	0.25
PCB	10		10
Arsenic	3	2–8	8
Barium	20	20–60	60
Cadmium			6
Chromium	25	25–75	75
Copper	25	25–50	50
Iron	17,000	17,000–25,000	25,000
Lead	40	40–60	60
Manganese	300	300–500	500
Mercury	1		1
Nickel	20	20–50	50
Phosphorus	420	420–650	650
Zinc	90	90–200	200

Adapted from United States Environmental Protection Agency (1977).

Table 7-8 Sediment Quality Guidelines of the Ontario Ministry of the Environment and Energy for Nutrients, Metals, and Organic Compounds (concentrations in µg/g dry weight unless otherwise noted)

	No effect level	Lowest effect level	Severe effect level[a]
Nutrients			
Total Kjeldahl nitrogen		550	4,800
Total organic carbon (%)		1	10
Total phosphorus		600	2,000
Metals			
Arsenic		6	33
Cadmium		0.6	10
Chromium		26	110
Copper		16	110
Iron (%)		2	4
Lead		31	250
Manganese		460	1,100
Mercury		0.2	2
Nickel		16	75
Zinc		120	820
Organic compounds			
Aldrin		0.002	8
BHC		0.003	12
Chlordane	0.005	0.007	6
DDT (total)		0.007	12
Dieldrin	0.0006	0.002	91
Endrin	0.0005	0.003	130
HCB	0.01	0.02	24
Hepoxide		0.005	5
Mirex		0.007	130
PCB (total)	0.01	0.07	530

[a] Numbers in this column are to be converted to bulk sediment values by multiplying by the actual total organic carbon concentration (to a maximum of 10%).

Adopted from Ontario Minstry of the Environment and Energy (1992).

sediments are classified as nonpolluted, moderately polluted, and heavily polluted (Table 7-7).

Recently, several environmental departments have developed additional chemical, concentration-based sediment quality criteria. The Ontario Ministry of the Environment and Energy (1992) has developed sediment guidelines for protecting and managing aquatic sediment quality in Ontario. The guidelines define three levels of ecotoxic effects and are based on the chronic, long-term effects of contaminants on benthic organisms (Table 7-8). These levels are:

1. *a no effect level,* at which no toxic effects have been observed on aquatic organisms;
2. *a lowest effect level,* indicating a level of sediment contamination that can be tolerated by the majority of benthic organisms; and
3. *a severe effect level,* indicating the level at which pronounced disturbance of a sediment-dwelling community can be expected.

The provincial sediment quality guidelines of OMEE (Table 7-8) supersede the open-water disposal guidelines and will provide the basis for all sediment evaluations (or potential lakefill materials to be placed in water) in Ontario.

In 1993, Reynoldson and Zarull made the first serious attempt to develop numerical, biological sediment guidelines. The authors proposed four approaches for developing numerical guidelines for comparison of observed and expected communities: (1) a simple scoring system; (2) percent difference in reference and test-site means (± the standard deviation for each taxa from the reference community); (3) calculation of X^2 values; and (4) comparison in multivariate space. In a more recent paper (Reynoldson et al., 1994), numeric criteria was developed for community structure and for toxicity tests based on a reference database from 335 clean sites in the North American Great Lakes. The guidelines are site-specific and use a predictive technique to select the appropriate value of a site.

However, many uncertainties remain regarding the scope of the contamination of aquatic sediments, how to distinguish a clean from a contaminated sediment, and what the legal basis is for regulating the sediment quality. Marcus (1991) recommended that in the immediate future "the best regulatory stance will be flexible, allowing for the testing of different evaluative techniques, the examination of various regulatory mechanisms, and the comparison of different cleanup techniques."

REFERENCES

Adeloju, S.B. and Bond, A.M., Influence of laboratory environment on the precision and accuracy of trace element analysis, *Anal. Chem.*, 57, 1728, 1985.

Ahlers, W.W., Reid, M.R., Kim, J.P., and Hunter, K.A., Contamination-free collection and handling protocols for trace elements in natural fresh waters, *Austr. J. Mar. Freshwater Res.*, 41, 713, 1990.

Anon, Ocean dumping: final regulations and citeria, *U.S. Federal Register*, 38, 1973.

Azcue, J.M., Metales en el medio ambiente, in *Metales en Sistemas Biologicos*, Mas, A. and Azcue, J.M., Eds., PPU, S.A., Barcelona, 1993, 163.

Beak Consultants Ltd., *Development of Sediment Quality Objectives*, Report prepared for the Ontario Ministry of Environment, Canada, 1987.

Boutron, C.F., A clean laboratory for ultralow concentration heavy metal analysis, *Fresenius Z. Anal. Chem.*, 337, 482, 1990.

Cheam, V. and Chau, A.S.Y., Analytical reference materials IV. Development and certification of the first Great Lakes sediment reference material for arsenic, selenium and mercury. *Analyst*, 109, 775, 1984.

Cheam, V., Aspila, K.I., and Chau, A.S.Y., Analytical reference materials VIII. Development and certification of a new Great Lakes sediment reference material for eight trace metals, *Sci. Total Environ.*, 87, 517, 1989.

Crepin, J. and Johnson, R.L., Soil Sampling for Environmental Assessment, in *Soil Sampling and Methods of Analysis*, Carter, M.R., Ed., Lewis Publishers, Chelsea, Michigan, 1993, 5.

Dutka, B.J., Jones, K., Kwan, K.K., Bailey, H., and McInnis, R., Use of microbial and toxicant screening tests for priority site solution of degraded areas in water bodies, *Water Res.*, 22, 503, 1988.

Fitchko, J., *Criteria for Contaminated Soil/Sediment Cleanup*, Puvan Publishing, Northbrook, Illinois, 1989.

Förstner, U., Ahlf, W., Calmano, W., and Kersten, M., Sediment criteria development — contributions from environmental geochemistry to water quality management, in *Sediments and Environmental Geochemistry*, Helling, D., Rothe, P., Förstner, U., and Stoffers, P., Eds., Springer-Verlag, Heidelberg, 1990, 312.

Giesy, J.P. and Hoke, R.A., Freshwater sediment quality criteria: toxicity bioassessment, in *Sediments: Chemistry and Toxicity of In-Place Pollutants*, Baudo, R., Giesy R., and Muntau, H., Eds., Lewis Publishers, Chelsea, Michigan, 1990, 265.

Håkanson, L., An ecological risk index for aquatic pollution control, a sedimentological approach, *Water Res.*, 14, 975, 1980.

Hamilton, E.I., Sediment — what is being measured?, *Mar. Pollut. Bull.*, 26, 58, 1993.

Hart, D.R., Fitchko, J., and McKee, P.M., *Development of Sediment Quality Guidelines, Phase II — Guideline Development*, Beak Consultants, Ltd., Brampton, Canada, 1988.

International Joint Commission, *Guidance on Characterization of Toxic Substances Problems in Areas of Concern in the Great Lakes Basin*, Report from the Surveillance Work Group, Windsor, 1987, 177.

International Joint Commission, *Procedures for the Assessment of Contaminated Sediment Problems in the Great Lakes Basin*, Report from the Surveillance Work Group, Windsor, 1988, 140.

Jenne, E.A. and Luoma, S.N., Forms of trace elements in soils, sediments and associated waters: an overview of their determination and biological availability, in *Biological Implications of Metals in the Environment*, Widung, R.R. and Drucker, H., Eds., U.S. Energy Res. Develop. Admin. Sys. Ser. 42, 1977, 110.

Keith, L.H., *Environmental Samples and Analyses: A Practical Guide*, Lewis Publishers, Chelsea, Michigan, 1991, 143.

Keith, L.H., Crummett, W., Deegan, J., Libby, R.A., Taylor, J.K., and Wentter, G., Principles of environmental analysis, *Anal. Chem.*, 55, 2210, 1983.

Kosta, L., Contamination as a limiting parameter in trace analysis, *Talanta*, 29, 985, 1982.

Kratochvil, B., Walker, D., and Taylor, J.K., Sampling for chemical analysis, *Anal. Chem.*, 56, 13, 1984.

Lawrence, L., Chau, A.S.Y., and Aspila, K.I., Analytical quality assurance: key to reliable environmental data, *Can. Res.*, Nov., 35, 1982.

Laxon, D.P.H. and Harrison, R.M., Cleaning methods for polythene containers prior to the determination of trace metals in freshwater samples, *Anal. Chem.*, 53, 345, 1981.

Marcus, W.A., Managing contaminated sediments in aquatic environments: identification, regulation, and remediation, *Environ. Law Report.*, 21, 20, 1991.

Mart, L., Minimization of accuracy risks in voltametric ultratrace determination of heavy metals in natural waters, *Talanta*, 29, 1035, 1982.

Massee, R., Maessen, F.J.M.J., and Goeij, J.J.M., Losses of silver, arsenic, cadmium, selenium and zinc traces from distilled water and artificial sea-water by sorption on various container surfaces, *Anal. Chimica Acta*, 127, 181, 1981.

Moody, J.R. and Linstrom, R.M., Selection and cleaning of plastic containers for storage of trace element samples, *Anal. Chem.*, 49, 2264, 1977.

Mudroch, A., Sarazin, L., Leaney-East, A., Lomas, T., and deBarros, C., *Report on the Progress of the Revision of the MOE Guidelines for Dredged Material Open Water Disposal, 1984/1985*, Environment Canada, Inland Waters Directorate, Environmental Contaminants Division, 15, 1988.

Müller, G., Schwermetalle in den sedimenteen des Rheins — Verānderungen seit 1971, *Umschan*, 79, 778, 1979.

Murphy, T.J., *Accuracy in Trace Analysis: Sampling, Sample Handling and Analysis*, Sepec. Public. 442, La Fleur, P.D., Ed., National Bureau of Standards, Washington, 1976, 509.

National Bureau of Standards (NBS), U.S. Department of Commerce, Gathersburg, Maryland, 1992.

National Research Council of Canada (NRCC), Division of Chemistry, *Marine Reference Materials and Standards*, Ottawa, 1988.

National Water Research Institute (NWRI), *Marine Reference Materials and Standards, Sediment Reference Materials*, Aquatic Quality Assurance Program, Research & Applications Branch, Environment Canada, Burlington, 1992.

Nriagu, J.O. and Azcue, J.M., Food contamination with arsenic in the environment, in *Food Contamination from Environmental Sources*, Vol. 23, Nriagu, J.O. and Simons, M.S., Eds., John Wiley & Sons, New York, 1990, 103.

Nriagu, J.O., Lawson, G., Wong, H., and Azcue, J.M., A protocol for minimizing contamination in the analysis of trace metals in Great Lakes waters, *J. Great Lakes Res.*, 19, 175, 1992.

Ontario Ministry of the Environment and Energy (OMEE), *Guidelines for the Protection and Management of Aquatic Sediment Quality in Ontario*, Ontario Ministry of the Environment and Energy, Toronto, 1992.

Persuad, D., Jaagumagi, R., and Hayton, A., *Development of Provincial Sediment Quality Guidelines*, Ontario Ministry of the Environment and Energy, Toronto, 19, 1989.

Peterson, R.G. and Calvin, L.D., Sampling, in *Methods of Soil Analysis. Part I. Physical and Mineralogical Methods*, Agronomy no. 9, 2nd ed., Klute, A., Ed., American Society of Agronomy, Madison, 33–52, 1986.

Rasberry, S.D. and Gills, T.E., The certification and use of standard reference materials, *Spectrochim. Acta*, 46, 1577, 1991.

Reynoldson, T.B. and Zarull, M.A., An approach to the development of biological sediment guidelines, in *Ecological Integrity and the Management of Ecosystems*, Woodley, S.J., Francis, G., and Kay, J. Eds., St. Lucie Press, Delray Beach, 1993.

Reynoldson, T.B., Bailey, R.C., Day, K.E. and Norris, R.H., Biological guidelines for freshwater sediment based on BEnthic Assessment of SedimenT (the BEAST) using a multivariate approach for predicting biological state, *Aust. J. Ecol.*, (in press), 1994.

Robertson, D.E., Role of contamination in trace element analysis in sea water, *Anal. Chem.*, 40, 1067, 1968.

Shea, D., Developing national sediment quality criteria — equilibrium partitioning of contaminants as a means of evaluating sediment quality criteria, *Environ. Sci. Technol.*, 22, 1256, 1988.

Standard Methods, American Public Health Association, American Water Works Association, and Water Environment Federation, *Standard Methods for the Examination of Water and Wastewater*, 18th ed., Washington, 1992.

Sturgeon, R. and Berman, S.S., Sampling and storage of natural water for trace metals, *CRC Critical Reviews in Analytical Chemistry*, 18, 209, 1987.

Tessier, A., Campbell, P.G.C., and Bisson, M., Sequential extraction procedure for the speciation of particulate trace metals, *Anal. Chem.*, 51, 844, 1979.

United States Environmental Protection Agency (U.S. EPA), *Guidelines for the Pollutional Classification of Great Lakes Harbour, Sediments*, U.S. Environmental Protection Agency, Region V, Chicago, 1977.

United States Environmental Protection Agency (U.S. EPA), *Background and Review Document on the Development of Sediment Criteria*, EPA/68-01-6388, JRB Associates, McLean, Virginia, 1984.

United States Environmental Protection Agency (U.S. EPA), *ASSESS Users Guide*, U.S. Environmental Protection Agency, Environmental Monitoring Systems Laboratory, EPA/600/8-91001, Las Vegas, 1991.

van de Guchte, C. and van Leeuwen, C.J., Sediment pollution, in *Manual on Aquatic Ecotoxicology*, de Kriuff, H.A.M., de Zwart, D., Ray, P.K., and Viswanathan, P.N., Eds., P.N. Kluwer Acad. Dordrecht, Netherlands, 1988.

Van Loon, J.C., *Selected Methods of Trace Metals Analysis. Biological and Environmental Samples*, John Wiley & Sons, New York, 1985.

Van Loon, J.C. and Barefoot, R.R., *Analytical Methods for Chemical Exploration*, Academic Press, San Diego, 1989, 344.

Webster, R. and Oliver, M.A., *Statistical Methods in Soil and Land Resource Survey,* Oxford University Press, Oxford, 1990, 315.

CHAPTER 8

Costs Associated with Sediment Sampling

The cost of sediment sampling and processing includes all expenditures for collecting sediments into proper containers, making measurements and handling samples in the field, transporting and storing samples, preparing samples, performing sediment tests and analyses, instituting quality control procedures, processing data, and preparing reports.

The cost of sediment sampling depends (in decreasing order) first, on the number of sampling locations; second, on the selected sampling procedures; third, on the number of samples to be collected and/or on the amount of sediment material required for analyses and experiments; and fourth, on sample handling and preservation procedures. The number of sampling locations and the selected sampling procedure greatly affect the cost of sample collection and are factors that, with regards to available funding, will contribute to deciding the feasibility of the project. Obviously, there is a difference between the cost for collecting a sediment sample in a small, shallow lake from a small vessel operated by two workers, and the cost for collecting a sediment sample in a large, deep lake or sea that requires a vessel of appropriate size and an experienced crew.

It should be expected that the cost of sediment sampling will vary from one project to another. However, there are several costs commonly encountered in sediment sampling programs, as follows:

- Transportation of scientific and technical personnel to the sampling location, and their accommodation and meals.
- Shipping equipment to and from the sampling location.
- Renting/leasing/purchasing and operating vessels, cars, trucks, planes, helicopters, snowmobiles, etc., required for transporting the sampling equipment to the sampling location.
- Preserving and shipping collected samples.
- Salaries for scientific and technical personnel.
- Hiring professional services, such as divers, or local personnel to help in the sampling program.

- Overheads, i.e., allowance to cover miscellaneous services and maintenance costs, etc.

Table 8-1 shows examples of costs associated with sediment sampling in four environmental studies. For comparison purposes, the four selected studies had similar scientific objectives (the study of environmental impacts of mining activities), but were characterized by very different field conditions. Moreover, study Nos. 1 and 2 were carried out by government agencies, study No. 3 by a university, and No. 4 by an environmental consulting company. The cost for individual items is expressed as a percent of the total cost of each sediment sampling program. The costs for sample preparation in the laboratory and physico-chemical analysis of collected samples — with necessary QA/QC — are not included in the table.

Study No. 1 in Table 8-1 was carried out in the Cariboo Range, British Columbia, Canada. The study was initiated to evaluate the effects of abandoned gold mine tailings on the adjacent terrestrial and aquatic ecosystems. The major objectives of the study were to (1) determine the distribution of major and trace elements in different environmental compartments, such as suspended and bottom sediments and sediment pore water at the site; (2) evaluate the effects of the major and trace elements in the bottom sediments on benthic community structure in a lake adjacent to the tailings; and (3) evaluate the tailings as a source of contaminants to the adjacent surface waters. Bottom sediments and sediment pore water were sampled by two scientists and one technician, within one week, at ten sampling stations in the 3-km long, 0.5-km wide lake adjacent to the tailings and at two stations in a control lake located about 15 km from the study site. Accommodation was provided for the scien-

Table 8-1 Costs Associated with Sediment Sampling

	Percentage of total cost			
Expenses	**Study no. 1**	**Study no. 2**	**Study no. 3**	**Study no. 4**
Shipping equipment	25	30	10	7.8
Transportation (to and from site)	24	40	47	31
Rental of cars, trucks, or boats	20	10	12	13.6
Accommodations and meals	17	5	12	4.1
Shipping samples	5.6	10	7	3.5
Professional services	a	a	a	35
Field supplies (chemicals, tapes, cables, etc.)	8.4	5	12[b]	5

[a] The salaries of all members of the sampling party were covered by government agencies or universities. Therefore, these are not included in the costs associated with the sediment sampling.

[b] Use of radioactive tracers.

tific and technical personnel at a motel conveniently located about 1 km from the sampling site. A small aluminum boat with a 5-hp engine used for the sediment sampling was lent by the local government agency. Personnel and equipment involved in the sediment sampling were transported by commercial airline (about 3,000 km distance one-way); a truck and a personal vehicle were rented at the airport.

Study No. 2 in Table 8-1 was carried out in the Northwest Territories, Canada. The study was initiated to assess the effects of abandoned gold mine tailings on the aquatic and terrestrial environment. The major objectives of the study were to (1) determine the physico-chemical character of the runoff from the tailings; (2) determine the geochemical composition of the tailings that were in contact with surface waters; and (3) evaluate leaching of different elements from the tailings and their accumulation in the bottom sediments in the adjacent lakes and streams. Bottom sediments were sampled by two scientists at five sampling stations in the lakes in the vicinity of the tailings (within about a 3-km distance) within one day. A Turbo-Beaver float plane was rented to transport the sampling personnel and all equipment to and from the 83-km-distant study site.

Study No. 3 in Table 8-1 was carried out in the Madeira River and its tributaries in Brazil. The study was initiated to evaluate the impact of gold mining activities in the Amazon River basin on the Madeira River ecosystem. The major objectives of the study were to (1) obtain data on the distribution of mercury and other heavy metals originating from the gold mining activities in the Amazon River basin in bottom and suspended sediments; (2) obtain information on chemical forms of mercury in sediments and sediment pore water; and (3) evaluate the potential net rates of methylation of mercury in surficial sediments. Bottom sediments were collected at twenty-three sampling stations in the Madeira River and its tributaries within a distance of about 240 km over a two-week period. The sampling, carried out by four scientists, was made possible through collaboration with various institutions and companies who provided cost-free accommodation, field laboratory space with freezers and fridges, a truck and a jeep with drivers, and a 5-m aluminum boat with an operator familiar with the sampling area. A detailed description of the sampling program for study Nos. 1 to 3 was given by Azcue et al. (1994).

Study No. 4 in Table 8-1 was carried out in Buttle Lake, British Columbia, Canada. This study was part of a larger project to assess the geochemical reactivity of mine tailings covered by natural sediments. The major objectives of the study were to (1) assess the geochemical reactivity of mine tailings covered by natural sediments; (2) study the effects on tailings reactivity induced by seasonal changes; and (3) determine the efficacy of short- and long-term subaqueous disposal of mine tailings. Buttle Lake is a large, deep lake (35 km long × 1 km wide × 80 m deep). Three stations across the lake were sampled for bottom sediments and sediment pore water. The sampling was carried out within ten days by two scientists and one technician. All sampling equipment was transported to the field in a rented truck. For this study, due to the depth of the lake, professional diving services were contracted in the local town. The details of this project were described by Rescan (1993).

In all four studies, the cost for transportation with added cost for the rental of cars, trucks, and boats, represented the greatest part of the total cost for the sediment sampling.

In planning and budgeting the sediment sampling program, it should be recognized that even well-planned activities can go astray, and that various problems can arise, such as sickness of the workers, loss of sample containers, poor-quality samples requiring additional sampling, etc. Contingencies for such incidents, as well as additional funds, should be considered during the planning stage. One factor that should always be considered in cost estimates is the weather and its caprices. Poor weather, for example storms, can interfere in sediment sampling, causing delays and unexpected expenses.

The cost of chemical, physical, and geotechnical analyses, bioassays, and other tests outlined in the project plan have to be estimated as accurately as possible and added to the estimated cost of sediment sampling. The costs of laboratory measurements can be obtained from private or governmental laboratories involved in the type of analyses required in the project. If the project budget exceeds available funds, alternatives will have to be considered, for example, cancelling or postponing the project or revising the project objectives and working plan.

REFERENCES

Azcue, J.M., Guimaraes, J.R.D., Mudroch, A., Mudroch, P., and Malm, O., Case studies, in *Handbook of Techniques of Aquatic Sediment Sampling*, 2nd ed., Mudroch, A. and MacKnight, S.D., Eds., Lewis Publishers, Chelsea, Michigan, 1994.

Rescan Environmental Services Ltd., *Subaqueous Tailings Disposal Study — Buttle Lake Summer Survey 1993*, Mine Environmental Neutral Drainage (MEND) Project 2.11, 1993.

INDEX

C

G

Q

R

S

T

U

V

W

Z